Electronic Projects from the Next Dimension: Paranormal Experiments for Hobbyists

Other Books of Interest by Newnes

ELECTRONIC CIRCUIT INVESTIGATOR SERIES
ROBERT J. DAVIS, *Digital and Computer Projects*
NEWTON C. BRAGA, *CMOS Projects and Experiments: Fun with the 4093 Integrated Circuit*

RELATED TITLES
ANDREW SINGMIN, *Beginning Electronics through Projects*
ANDREW SINGMIN, *Practical Audio Amplifier Circuit Projects*
ANDREW SINGMIN, *Beginning Digital Electronics through Projects*

Electronic Projects from the Next Dimension: Paranormal Experiments for Hobbyists

Newton C. Braga

Newnes

Boston Oxford Auckland Johannesburg Melbourne New Delhi

Newnes is an imprint of Butterworth–Heinemann.

Copyright © 2001 by Butterworth–Heinemann

 A member of the Reed Elsevier group

All rights reserved.

Although great care has been taken in the description of devices and experiments, the author wishes to alert readers to the dangers of using them without proper installation and due caution. Anyone making use of this information assumes all risk and liability arising from such use.

No part of this publication may be reproduced, stored in a retrieval system, or transmitted in any form or by any means, electronic, mechanical, photocopying, recording, or otherwise, without the prior written permission of the publisher.

 Recognizing the importance of preserving what has been written, Butterworth–Heinemann prints its books on acid-free paper whenever possible.

 Butterworth–Heinemann supports the efforts of American Forests and the Global ReLeaf program in its campaign for the betterment of trees, forests, and our environment.

ISBN 0-7506-7305-2

British Library Cataloguing-in-Publication Data
A catalogue record for this book is available from the British Library.

The publisher offers special discounts on bulk orders of this book.
For information, please contact:

Manager of Special Sales
Butterworth–Heinemann
225 Wildwood Avenue
Woburn, MA 01801-2041
Tel: 781-904-2500
Fax: 781-904-2620

For information on all Newnes publications available, contact our World Wide Web home page at: http://www.newnespress.com

10 9 8 7 6 5 4 3 2 1

Printed in the United States of America

Contents

Preface ix

Dedication xi

About the Author xii

Part 1: Instrumental Transcommunication 1
- 1.1 Principles and History. 1
- 1.2 Instrumental Transcommunication 1
- 1.3 Stochastic Resonance 2
- 1.4 Chronology of Transcommunication 8
- 1.5 Practical Circuits. 11
- 1.6 Observations about the Research 11
- 1.7 Circuits and Devices. 12
- 1.8 Materials Needed To Begin 12
- 1.9 Performing the Experiments 15
- 1.10 Interpretation. 16
- 1.11 Starting Up 16
- 1.12 Using a Tape Recorder to Pick Up "The Voices". 16
- 1.13 White Noise Generators 19
  - Project 1 White Noise Generator I 19
  - Project 2 White Noise Generator II. 24
  - Project 3 White Noise Generator Using an IC 27
- 1.14 Experimenting with White Noise Generators. 30
  - Project 4 Sound Filter 32
  - Project 5 Noise Filter 35
  - Project 6 60 Hz Filter (Hum Filter). 38
  - Project 7 Low-Impedance Preamplifier 41
- 1.15 Ultrasonic Sources 44
  - Project 8 Low-Power Ultrasonic Source. 45
  - Project 9 High-Power Modulated Ultrasonic Source 48
- 1.16 Experiments 52
- 1.17 Picking Up Sounds from the Earth 56
  - Project 10 Noise Modulated Transmitter 58
  - Project 11 Wireless Noise Generator
    (Medium Wave or Shortwave) 61

Part 2: Experimenting with Images 67

2.1 The Basic Idea..68
2.2 Experimenting with EIP 70
2.3 Scanning, Frames, and the TV Image......................73
2.4 Practical Circuits.......................................76
    Project 12 Wireless Sparkling Image Generator.................76
    Project 13 TV White Noise Generator 77
    Project 14 Horizontal Bar Generator 82
    Project 15 Video Inverter 87
2.5 Suggested Experiments....................................92
2.6 Combined Sound and Image Projects........................98
    Project 16 Light-to-Sound Converter 98
    Project 17 Sound-to-Light Converter 103
    Project 18 Brontophonic Sound 106
    Project 19 Ultrasonic Sound Converter 114
    Project 20 Laser Image Generator 119
    Project 21 Magnetic Field Generator 122
2.7 Kirlian Photography 128
    Project 22 High-Voltage Generator (Kirlian Machine I) 135
    Project 23 High-Voltage Generator (Kirlian Machine II).....146
    Project 24 High-Voltage, High-Power Generator
               (Kirlian Machine III) 151

Part 3: Experimenting with Paranormal Skills.................. 159

3.1 Paranormal Skills 160
3.2 Paranormal Phenomena 161
3.3 Sony's Seven-Year Paranormal Research Effort 168
3.4 Biofeedback Experiments and Projects 169
3.5 Experiments 173
    Project 25 Temperature Change Monitor 173
    Project 26 Temperature-Controlled Oscillator 177
    Project 27 Light and Dark Controlled Oscillator 180
    Project 28 Polygraph......................................186
    Project 29 Hypnotic LEDs 190
    Project 30 Electro-Shock Generator 195
    Project 31 Third Eye......................................200
    Project 32 Hypnotic Glasses...............................205
    Project 33 Bioprobe 211
    Project 34 Paranormal Electroscope 214
3.6 Paranormal Experiments with Light.......................218
    Project 35 Simple Light Detector 218
    Project 36 Psycholamps 223

	Project 37 Electronic Candle	229
3.7	ESP and PK Experiments	234
	Project 38 Random Number Generator	234
	Project 39 Binary Random Number Generator	240
3.8	UFOs and Ghosts	245
	Project 40 UFO Detector	245
	Project 41 Ghost Finder	250
3.9	Other Paranormal Experiments	255
3.10	The Computer	256

Bibliography **260**

Preface

*There are more things in heaven and earth, Horatio
Than are dreamt of in your philosophy.*

William Shakespeare (Hamlet)

Shakespeare was confident when writing those words and, even in our time, despite the fantastic progress of "official science," there are many more unexplained things in our world than our philosophy can dream.

Many people believe that "official science" is not relevant to "paranormal phenomena," leaving the subject to mystical groups, religious groups, amateur scientists, and others. They are wrong.

Each day, paranormal phenomena become more evident, and we cannot deny their existence. Little by little, these phenomena are being recognized as legitimate targets of official science.

Important research organizations such as universities, government organizations, and others are forming groups to work in what is referred to as *paranormal sciences,* including ESP (extra-sensory perception), EVP (electronic voice phenomena), ITC (instrumental transcommunication), poltergeists, telepathy, UFOs, and so on, in a clear admission that they believe in an old adage of my country, "Where exists smoke also exists fire."

Today, the groups are multiplying, and many are devoted to specific areas of research with paranormal phenomena. Most of these groups are motivated by a strong sense of curiosity and are trying to develop new experiments and theories. In many cases, they are trying to adapt the results of their work to fit their beliefs (religious or philosophical).

As the fundamentals of paranormal sciences are not included in the standard curriculum of our schools, anyone who wants to perform his own experiments will discover that there is a large barrier to traverse. There are few sources of information about scientific techniques and procedures that are suitable for practical experiments. There are no "official" methodologies to be used in a specific endeavor with those kinds of phenomena. And, what is more important to us, descriptions of practical instruments and devices that can be used for this research are rare. The consequences for readers who want to conduct work in this field are easy to see:

- A great deal of improvisation is required.
- Because of an absence of scientific information, it is difficult to separate what is real from what is false.
- There is a great lack of knowledge in using high-technology devices—and even common home appliances such as computers, microprocessors, and basic electronics.

Notice that when we use the term *high technology,* we are referring to electronics. This science offers vast resources to researchers in all scientific fields, which of course is not limited to paranormal phenomena. (What lab doesn't perform measurements with electronic instruments?)

In the process of giving support to groups of researchers, the author has found that many practical (and sometimes simple) electronic circuits can be used when making serious experiments involving paranormal phenomena. And some of the circuits are indeed simple—many can be assembled even by amateurs in a single weekend.

This book is a collection of circuits that the author has recommended to many researchers, curious individuals, and hobbyists during more than 20 years of work as technical director and author for two important electronics magazines in Brazil. Many of the circuits have been described in these magazines with objectives other than paranormal research, but the author has found it easy to make modifications for this task, and those modifications have been included in the projects included in this book.

To make it easy for the "common reader and researcher" to assemble the projects described in this book, the circuits have been simple and use only common parts. Many can even be found in old non-functioning devices such as radios, amplifiers, TV sets, etc. The circuits are not critical in design, which allows the reader to try his own modifications for increasing performance and/or changing the basic aim. In addition, the author invites readers to make alterations in the circuits (at indicated points) to broaden the range of applications.

The author also has included basic technical information about each project. Explaining how each circuit works makes it easier for the researcher to conduct the experiments correctly and safely. It also makes it easier to separate incorrect conclusions about what is happening from reality.

Finally, the author wants to remind readers that he intends only to give them the tools to conduct research work with paranormal phenomena. The results—explanations and theories about the nature of the phenomena—are left to the reader. But the author does not want to be disengaged from such matters. He invites readers to write him and tell about their results. Your discoveries may even point the direction for a new book.

Newton C. Braga

Dedication

To my great friend, Dr. Max Berezovsky, a great researcher of Paranormal Phenomena. Without the contribution of his vast knowledge about this subject, this book would not have been possible.

About the Author

Mr. Braga was born in São Paulo, Brazil, in 1946. He received his professional training at São Paulo University (USP). His activities in electronics began when he was 13 years old, at which time he began to write articles for Brazilian magazines. At age 18, he had his own column in the Brazilian edition of *Popular Electronics,* where he introduced the concept of "electronics for youngsters."

In 1976, he became technical director of the most important electronics magazine in South America, *Revista Saber Eletrônica* (published in Brazil, Argentina, and Mexico). He also has been the director of other magazines published by the same company, including *Eletrônica Total.* During this time, Mr. Braga has published more than 60 books about electronics, computers, and electricity, and thousands of articles and electronic projects in magazines all over the world (U.S.A., France, Spain, Portugal, Argentina, Mexico, et al.). More than 2,000,000 copies of his books have been sold throughout Latin America and Europe.

The author also teaches electronics and physics and is engaged in educational projects in his home country of Brazil. These projects include the introduction of electronics in secondary schools and professional training of workers who need enhanced knowledge in the field of electronics. The author now lives in Guarulhos (near São Paulo) and is married, with a 10-year-old son.

Part 1
Instrumental Transcommunication (Beyond Death with Electronics)

1.1 Principles and History

Nicola Tesla, the great inventor, believed that the newly discovered *radio* could be used to contact the spirits. Over many years, he and other famous inventors, including Marconi and Edison, searched for a way to "tune" their signals, upgrade their circuits, or otherwise make changes to achieve the new, fantastic purpose. They were not completely successful with their intent, although Edison described some strange signals picked up by some of his apparatus. But the basic question remained, "Can electronics be used to contact spirits?"

The movie "The Grass Harp" opened the door to a relatively unknown phenomenon that can be a key to the idea of an apparatus to communicate with the dead. The movie tells the history of a little girl who could hear the voices of the dead in plant leaves shaken by the wind. This historical outline explores a real phenomenon that is of increasing interest now that electronic resources are more readily available.

1.2 Instrumental Transcommunication

Instrumental transcommunication (ITC) is the name of this new para-science. The basic idea is based on enlisting the aid of electronic devices to record voices from the beyond that appear when the sound of the leaves being shaken by the wind (white noise) are simulated by electronic circuits.

The important concept to keep in mind is that "The Grass Harp" explores a real fact and that experiments for hearing the dead can be conducted by anyone, including you, using common electronic circuits and equipment.

The age for this interesting research began when the *electronic voice phenomenon (EVP)* was discovered. The EVP is part of man's endeavor to establish contact with dead or other dimensions' beings using electronic instruments.

The discovery was made in 1959 by Friedrich Jüngerson, a Swedish researcher, and it started with a wave of interest in this field. Jüngerson was using a common tape recorder to record birds' songs when he discovered mysterious voices in the recordings. Mixed with the background sound (the wind shaking leaves as in "The Grass Harp"), it was possible to hear conversations of people who were not visible in the place where the recordings were made, and there were also other mysterious sounds. The electronic voices were thought to be as-

Edison

Raudive

sociated with beings who lived many years before in that place, or with beings of other dimensions.

The basic idea is that the background noise acts as a carrier or support and can be modulated by any other kind of signal—not necessarily electric fields (or *energies*, as commonly termed by researchers in this field) that are present in free space.

Without background noise, the signals are too weak to be detectable, but, in presence of a noise, their levels increase beyond the threshold of detection and become clear—sometimes loud.

This phenomenon, also called *stochastic resonance,* is well known in physics and telecommunications.

1.3 Stochastic Resonance

Stochastic resonance occurs when a modulated signal that is too weak to be heard or otherwise detected under normal conditions becomes detectable due to a resonance phenomenon produced between this weak deterministic signal and the signal with no definite frequency—the stochastic noise. Figure 1 shows what happens. The weak signal is covered by the noise under normal conditions and is not detectable. But the same noise can be added to the weak signal, increasing its amplitude. In such a case, the undetectable signal rises to a detectable level.

In the case of the experiment made by Jüngerson, the background noise (wind shaking the leaves) provided exactly the support necessary to make the undetectable voices become detectable and appear when the tape was heard, as Fig. 2 suggests.

Researchers can't explain why the voices are not audible in the ambient where they are produced but appear in the playback. Possibly the presence of an electric field is necessary for the manifestation of this phenomenon. This also would explain the apparent need for some kind of recording instrument in all the experiments involving EVP.

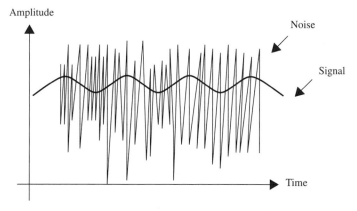

(a) The signal is covered by the noise.

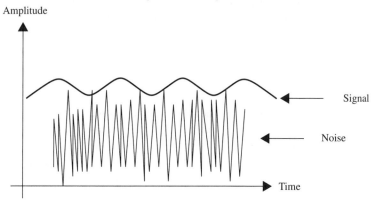

(b) The signal appears due to power added by the noise.

Figure 1 Stochastic resonance.

After Jüngerson's discovery, many other researchers began with a series of experiments, trying to pick up voices from the beyond. They first used common tape recorders as Jüngerson did. Later, other, more complex devices were implemented, including computers and TVs. The most important of these researchers was certainly Dr. Konstantin Raudive. He died in 1974 but left a number of works aimed at convincing humans that contact between the dead and the living is possible.

Dr. Raudive used not only common recorders to make his experiments but the complete range of equipment with which the principle of the stochastic resonance could be applied. Cassette tape recorders, video recorders, radio receivers, CD players, and even computers have been used in the experiments, beginning a new line of research that now also involves images.

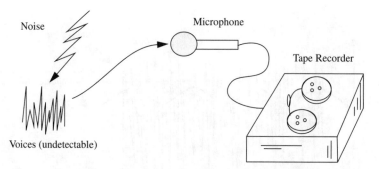

Figure 2 Voices appear in the playback, boosted by the background noise.

To produce the phenomenon of stochastic resonance, it is necessary to use some form of support or carrier to increase the amplitude of the signal to be detected. The experiments and devices used in transcommunications are all based on some kind of a noise source, with the most common being *white noise*, although other types of noise (e.g., *pink noise*) can be used.

1.3.1 White Noise

Physicists define *white noise* as a random sound within prescribed volume and tonal parameters. A more complex way to define white noise is as a signal with equal power per frequency unit (hertz) over a specified frequency band.

Another, more technical, definition is, "White noise is noise with an autocorrelation function zero everywhere but at 0, also called *Johnson noise*. It has a constant frequency spectrum."

The main sources of natural background noise are molecular-level agitation of matter via temperature changes, and natural electric discharges in the atmosphere. When we put a shell over our ear to hear "sea" noise, we are adding an acoustic amplifier to our natural hearing organs. The shell increases the white noise level produced by the thermal agitation of the air molecules to a value above the audible threshold. This is an audible white noise.

The same concept can be applied to electric signals spreading the frequency band to infinite values. This means that high-frequency circuits such as those found in radios can also be used to perform experiments in transcommunication (see Fig. 3).

The natural atmospheric discharges fill the electromagnetic spectrum with white noise that appears when we tune a radio receiver between stations as audible white noise (also known as *hiss*) or in a TV as visible "snow" when it is tuned to a free channel.

Many electronic components (e.g., resistors, diodes, transistors) can be used to produce white noise covering a large band of frequencies. These devices will be used as the basis of white noise generators in many of our projects.

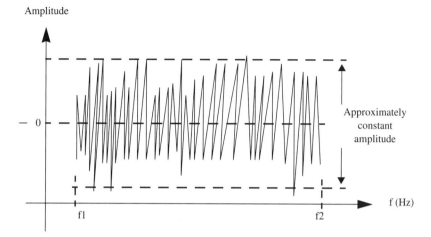

Figure 3 White noise spectrum between f1 and f2.

1.3.2 Pink Noise

Another kind of noise that can be used in projects involving transcommunication is *pink noise*. The difference is that pink noise has an amplitude that decreases along the frequency spectrum as shown in Fig. 4.

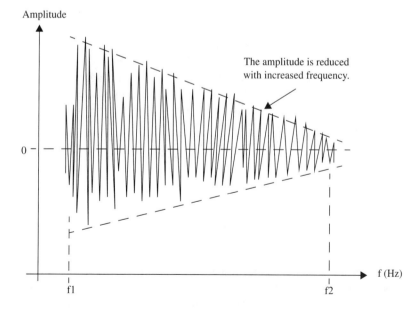

Figure 4 Pink noise spectrum between f1 and f2.

Noise can also be present in imaging processes. The snowy image on a TV is the result of noise present in the received signal or in the circuit of the TV set. The noise in a image can also be used in many experiments to detect "images from the beyond."

In the 1970s, the American pioneer George W. Meek began researching electronic voice phenomena with a new aim: *to achieve prolonged two-way communication rather than simply receiving short and isolated phrases as found in previous experiments.* He created a device, named the *Spiricon,* designed to maintain communication with the place from whence the voices were coming. He believed that the voices "came from beings residing in astral planes." The apparatus was basically a high-frequency oscillator, with the first version operating at 300 MHz and later versions at 1,200 MHz. The signal was used to fill the experiment room and was detected by a special circuit (a demodulator). The results showed that a solution to the problem did not lie in using high-frequency signals.

William J. O'Neil obtained the best results by lowering the frequencies used in the experiments. The new apparatus operated at 29 MHz.

Shortly afterward, the next generation of researchers appeared, and in 1982 an event at the National Press Club in Washington, D.C., attracted more than 50 reporters and researchers from many countries. Immediately, the news about the experiments spread all over the world, and a host of researchers began their experiments in countries such as Brazil and Russia, and throughout Europe and Asia.

Names including the German psychic Klaus Schreiber; Brazilians Max Berezovsky, Sonia Rinaldi, Hilda Hilst, and Ernani Guimarães; and many others made significant contributions and must be included among the important researchers in this field. The author, in several contacts with Dr. Max Berezovsky, learned much about the necessary equipment to undertake these experiments, creating many of the circuits found in this book from his suggestions.

We must also include the Luxembourg team, Jules and Maggie Harsch-Fishback, in our list of researchers who reported positive results in experiments involving ITC, and Ken Webster, who described how he received messages from the "other side" on his computer.

It is also important to remember that the first attempt in this field seems to be that of Jonathan Koons, in 1852, but the blueprints of his "machine" have never been found.

Sometime after the above-mentioned experiments were conducted, the name *instrumental transcommunication (ITC)* was first suggested for this parascience.

The origin of the "voices" or "images" is an interesting point of discussion. As described above, the assumption from which the first researchers began their studies was that the messages (voices or/and images) come from the dead, but other possibilities soon had to be considered. One of them is that the voices came from other-dimensional beings. These beings would live in a "parallel universe" with several degrees or levels or planes (astral planes). The contact with us, according to the researchers, is constant but sometimes not perceived by us. "They are here, but we can't see them," explain the researchers. Many believe that they act as "angels," protecting or acting on our lives and destinations without our observation.

Another theory says that the messages are generated inside our brain, and the background noise could trigger some internal feedback processes, making it detectable. Subconscious minds or even external signals can be transferred to the equipment and detected, appearing superimposed to the sounds and images, giving to the persons the impression that the voices came from an external source and not from their brain, as suggested in Fig. 5.

In this case, the brain will act as an "interface," transferring some kind of unknown and undetectable information from somewhere to an electronic device, making it visible or audible, or placing it on a magnetic tape or other medium such as a computer printer. This could explain why experts recommend that experimenters be in a special mind state when making these experiments!

Notice that this cannot be attributed to a psychic disorder in which the sounds are imagined by one person. The noise will help the brain to "tune" the information coming from outside and "add" them to the ambient noise in such a manner that all persons near the "receiver" can also hear them.

There are also people who believe that the messages come from a universal "data file" where all we say and do is stored and can be accessed anytime thereafter. The electronic equipment is used as a channel or medium to tune in those data, although the use of recorders in this manner precludes the possibility of choosing exactly the information we want.

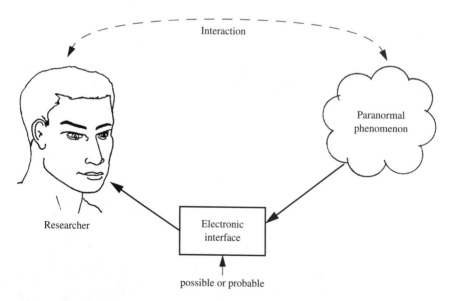

Figure 5 Action of the research on the paranormal phenomenon and an electronic interface.

It is like a situation in which we have discovered the radio receiver but not the way to tune to a desired station. What we pick up and when is not totally controlled yet. Researchers are still trying to find how to do that.

It is not for the author to decide what is the real or the correct answer to all these issues, as the researchers have reached no conclusions, and many doubts and questions about this subject are still unanswered. As in all science, there is much noise created between people who seek reality and those who want to use ITC for nonscientific purposes. Serious research will separate the noise from the signal—we hope.

Because this book is directed toward individuals who want to conduct practical experiments using available equipment and circuits, rather than constructing advanced theories, considerations about the nature of the phenomenon are not included. For present purposes, we can only suggest experiments and equipment based on existing theories and the work of well known researchers. A list of books about the subject will be given elsewhere in this book. There are also many sites on the Internet where the reader can find much useful information, not only about the researchers but also about their works. Consequently, a reader who is attracted to this unusual aspect of electronics and wants to make his own experiments will find some practical circuits herein.

1.4 Chronology of Transcommunication

- 1852—Jonathan Koon described a machine to communicate with the spirits, but the blueprints were never found.
- 1888—Nicola Tesla (U.S.A.) invented the induction motor, using it both to power transmissions and for radio wave generation. At the time, he suggested that the radio could be used to communicate with the dead.
- 1893—Father Landell de Moura (Brazil), who made experiments with radio transmissions before Marconi and patented them in U.S.A., reported the possibility of communications with the dead using radio waves. Some people said he had built a prototype, but documents or notes never were found.

Father Landell de Moura, the Brazilian researcher who made experiments with radio waves before Marconi.

- 1925—Oscar D Àrgonnel (Brazil) published the book, *Vozes do Além por Telefone (Voices from Beyond by Telephone)*, in which he described contacts with beings of other dimensions by telephone.
- 1923–1928—Cornélio Pires (Brazil) worked on an electronic apparatus intended to communicate with the spirits.
- 1928—Thomas Alva Edison (U.S.A.) introduced equipment he hoped could be used to communicate with the dead. The device used a chemical apparatus with potassium permanganate.
- 1933—Pròspero Papagesse (Brazil) described in a magazine the "Mediunic Electric Apparatus," an equipment intended to allow communications with the dead.
- 1936—Attliz von Szalay made experiments with a Packard-Bell recorder cutter and player, trying to pick up paranormal sounds on a phonograph record.
- 1947—Attilz von Szalay used a Sears & Roebuck wire recorder to register better quality voices.
- 1956—Based on von Szalay's experiments, Raymond Bayless recorded paranormal voices and wrote an article for the *Journal of the American Society for Psychical Research* in 1959.
- 1959—Friedrich Jüngerson (Sweden), when recording bird songs, discovered paranormal voices on playback. Four years later (1963), he called an international press conference to announce what he had discovered. One year later, in Stockholm, he published the book, *Voices from the Universe*.

Freidrich Jüngerson discovered the voice when recording bird songs.

- 1960–1970—Scott Rogo and Raymond Bayless (U.S.A.) conducted extensive literature research and published the book, *Phone Calls from the Dead*.
- 1964—von Szalay recorded voices of his deceased relatives on tape.
- 1965—Konstantin Raudive (Latvia) visited Jüngerson and concluded that the phenomenon was genuine. After this encounter, he started with his own experiments in Bad Krozingen (Germany).

- 1967—Theodore Rudolph developed a goniometer for Raudive's experiments.
- 1968—Father Leo Schmidt, in Switzerland, made experiments by taping voices. His book, *Wen Die Toten Reden (When the Dead Speak)*, was published in 1976, shortly after his death.
- 1970—Dr. Max Berezovsky began experiments with sounds, trying to pick up the voices.
- 1971—*Breakthrough, an Amazing Experiment in Electronic Communication with the Dead*, a book by Marcelo Bacci and co-workers in Grosseto, Italy, was published by Colin Smythe Ltd. (England), formed from expanded English translations of Raudive's book.
- 1971—Paul Jones, G. W. Meek, and Hans Heckman (U.S.A.) started a research program, trying to create a two-way voice communication system.
- 1972—Peter Bander (England) wrote the book, *Voices from the Tapes: Recording from the Other World*.
- 1973—Joseph and Michael Lamoreaux recorded paranormal voices after reading Raudive's book.
- 1975—William Addams Welch, a scriptwriter and playwright, authored "Talks with the Dead."
- 1978—William J. O'Neil, using a modified single side-band (SSB) radio, had brief but evidential contact with the dead.
- 1980s—Many researchers in many parts of the world had pictures of the "dead" appear sporadically on their TVs.
- 1980—Dr. Max Berezovsky (Brazil) began experiments with images using a video camera.
- 1980-1981—Manfred Borden (Germany) obtained unsolicited computer print-outs of beings from "other dimensions."
- 1981-1983—Manfred Boden had contacts with communicators by telephone.
- 1982-1988—Hans Otto Koenig (Germany) created an electronic equipment using very low frequency oscillators, ultra-violet and infrared lights, and other resources.
- 1984-1985—Kenneth Webster (England), using different computers, received 250 communications from a person who lived in the 16th century.
- 1985-1988—Jules and Maggie Harsh-Fischback (Luxembourg) created two electronic systems to be used in instrumental transcommunication.
- 1987-1988—Jules and Maggie Harsh-Fischback made contact with beings from other dimensions using the computer.
- 1985—Klaus Schreiber and Martin Wenzel (Germany) received images of dead persons on TV picture tubes using optoelectronic feedback systems. Rainer Hobbe of Radio Luxembourg wrote a book about it and used the subject for a TV documentary film.
- 1987—Jules and Maggie Harsh-Fisch (Luxembourg) received TV picture sequences from communicators from beyond.

1.5 Practical Circuits

The purpose of this part of the book is to describe experiments and projects involving transcommunication based in the stochastic resonance phenomenon. This means that the circuits and experiments described next are basically formed using noise sources added to some kind of configuration to increase performance, to alter the frequency band, or to produce other effects.

Starting with electronic voice phenomenon (EVP) experiments, the circuits and projects will increase in complexity. At the end, we will describe some applications involving work with images and a computer.

To the reader who is just starting with these experiments, is important to remember that the main attraction of the EVP, and also the electronic image phenomenon (EIP), is the fact that they can be reproduced by anyone. Both phenomena are studied via ITC.

1.6 Observations about the Research

We are working with a phenomenon that is not well known, and great care is important to achieve good results. Many factors are involved in successful ITC work.

The researchers recommend special care with human factors such as the attitudes, beliefs, and thoughts of the unknown beings to be contacted. This is the "mystic" aspect of the research, and each reader is free to adopt the appropriate conduct according to his beliefs. Others factors include technical details of the equipment and even computer programs that can be used by the researchers and (as some believe) by the beings on "the other side." There are also spiritual factors that can influence the experiments, and these also depend on each reader's beliefs.

Another point to observe is related to technical considerations when describing the results of an experiment. We are working with an unknown phenomenon in many aspects. The natural tendency of researchers is to use terms adopted by "official" science that have a very well stabilized meaning to describe these unknown phenomena. This is sometimes a very critical consideration when working with "nonacademic" research.

The most common case is the use of the word *energy* to describe every manifestation detectable in our world. "Official science" doesn't like this usage. The incorrect use of technical terms when describing an unknown phenomenon is one of the sources of disbelief when researchers try to describe the results of their experiments or to bring forth a new theory using improper words.

Energy, work, force, and *field* are terms with well defined meanings, and their use to describe the unknown causes a certain discomfort in academia. Researchers can avoid problems by trying to use the correct term or by creating new ones when necessary. Never use a scientific term to describe a phenomenon where it has no meaning or where you are not sure if it is proper usage. Take particular care when using words such as *energy* or *force*.

1.7 Circuits and Devices

From this point on, the reader will find a large assortment of circuits, equipment, and configurations that can be easily built from common electronic parts and perhaps even components found in old nonfunctioning devices such as radios, TV sets, amplifiers, and so on.

The intent is to give researchers technical support for using electronic equipment (commercial and home-built) and when setting up the experiments. Suggestions for alterations or changes in the practical circuits are given in many cases, furnishing many combinations of performances and letting the reader create as many experiments as desired.

How the circuits and devices work is another important point to consider when explaining the results to others. Transcommunications projects are basically the ones that produce some kind of noise to serve as support or carrier for the unknown signals to be picked up. Many configurations will be given, and with them the necessary information to adapt, transform, or upgrade according to the needs of each researcher.

1.8 Materials Needed To Begin

Recording voices from the beyond and performing experiments with the EVP doesn't require expensive equipment or hard-to-find instruments or parts. Starting with common commercial equipment, we can add some circuits to improve performance and help the researcher to find the voices in the tapes. So, before presenting our circuits, it is important to provide the reader with the following list of basic materials that will be required.

- Tape recorder (cassette or reel-to-reel)
- Tapes (low-noise C60 or C90 for cassette recorders)
- Good quality microphone (omnidirectional electric condenser, if possible)
- White or pink noise generator
- A low/medium-power audio amplifier
- A journal for notes
- High-quality headphones or processing equipment

Some comments follow about this basic equipment.

1.8.1 Choosing a Tape Recorder

It has been found that the voices are not heard by the persons in the ambient filled with noise but appear only in the playback when the noise is recorded. This means that the experiments described in the following pages start from information (voices) recorded on a magnetic tape. So, the reader needs, as first and basic equipment for the EVP lab, a tape recorder.

Old tape recorders such as the ones that use cassette tapes (as shown in Fig. 6) are recommended. It is important to use a recorder with a counter. This will help you to find any important part of a recording by the position in the tape. Microrecorders that use minicassette tapes can also be used, but the recording time is reduced, as is the sound quality.

You can also use professional-level recorders or reel-to-reel machines, but they are not easily found today. If you are lucky, you can find a suitable tape recorder in a pawn shop or electrical service shop and buy it at a reduced price.

Digital tape recorders can also be used, but there some limitations in their use. The first occurs because, when digitizing the signals, part of the noise and information can be lost. The second factor is that common digital recorders are not intended for long-term recording. They can only record a few minutes of sounds, which is not enough for serious work when trying to detect the voices.

A computer can be used as a tape recorder, but there are also some limitations. One limitation is that a sound file needs a large amount of space on your hard drive, and even if you transfer the files to diskettes, they are not well suited to long-term recording. As suggested in the end of this section ("Using the Computer"), the PC is a valuable tool for analyzing recordings from an analog tape, transferring them to digital files, and using software resources for analysis.

1.8.2 The Tapes

It is important to use new tapes. The residual sound recorded in a tape can appear as background sound in some cases, masking the experiments. Remember that magnetic tapes can lose the information recorded on them in a few years. C60 or C90 cassette tapes are recommended. Avoid the presence of magnets near the tapes, and never leave them in hot places such as inside a car on a summer day.

1.8.3 Headphones

It is not easy to hear the voices without the aid of a good set of headphones. In some cases, the recorder doesn't have an output to connect this device. In this

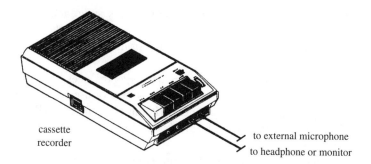

Figure 6 An old-fashioned tape recorder is best for experiments with EVP.

case, it is necessary to add a jack plug to the recorder. The output for an external earphone (monitor) is also important to plug it to an external amplifier. This allows you to share the recordings with others by playing them over a loudspeaker.

1.8.4 Microphones

Simple cassette recorders have an internal microphone. Normally, this microphone is not sensitive enough to perform the experiments with good results. The white noise used as support in the experiments has an important part of its energy concentrated in the high-frequency band (treble), and the internal microphone doesn't pick up those frequencies well. The use of a high-quality external microphone is important to get better results when performing these experiments.

1.8.5 White or Pink Noise Generator

As described before, the EVP is based on the support of undetectable unknown vibrations by some kind of noise such as white or pink noise. The simplest noise source is a running tap, a fan, the wind, or some operating electric appliance.

However, the more advanced experimenter can use electronic circuits to produce noise. Such sources are described in the following pages. The advantage of the use of an electronic noise source is that you can control it to an extent that is impossible with other kinds of sources. Dynamic range, intensity, and other characteristics can be controlled with precision or altered by changing parts of the circuit.

1.8.6 Audio Amplifier

Some noise generators don't include an audio output stage with enough power to drive a loudspeaker. The noise generator is, in general, only a simple circuit that produces a very low-power signal to be injected into low signal circuits such as the input of tape recorders, mixers, amplifiers, etc. Therefore, if the reader wants to fill an ambient with noise from this kind of noise source, it is necessary to use an audio amplifier.

The noise generator is plugged to the input of this audio amplifier, and the noise is reproduced by its loudspeaker, filling the ambient as shown in Fig. 7. Any audio amplifier with output power ranging from 0.5 to 20 W is suitable for these experiments.

The reader can build (from kits or components as the ones suggested in the next pages) small- or medium-power audio amplifiers or use commercial units, including any portable or fixed equipment with an audio input (AUX or other). This amplifier can also be used in the playback processing and to perform many other experiments with the EVP.

If the audio amplifier includes a tone control or a graphic equalizer, the experimenter can use it to alter the dynamic range of the noise.

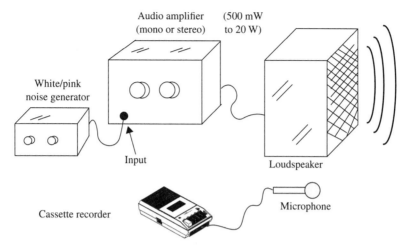

Figure 7 Using an audio amplifier in the experiments.

1.9 Performing the Experiments

Locate a silent place to make your experiments. Any undesirable sound that can be picked up by the microphone or other sensor and doesn't correspond to the voices can confuse the experimenter. Televisions and radios in adjacent rooms can produce undesirable interference to appear in your tapes when playing them back.

Put a tape in your recorder and turn on your noise source to fill the ambient with sound. Let the tape run.

You can program sessions lasting between 10 and 15 minutes. It is useful to begin each session by speaking something that will identify the recording sessions, such as the date and the time.

Some researchers begin with some questions to the voices such as "Who are you?" "Is anyone here?" or "Who exactly are you?"

The next step is to examine the tapes by playing them back and trying to find the voices mixed with the noise. Reset the counter to 000 and label your tape with the date the session was taped.

Put on the headphones and find the best volume for playback. In your notes, document any different sounds you find, giving the counter position to find it easily again, if necessary.

Researchers say that, in the beginning, the voices will most likely be a whisper, but as you advance with your experiments, they will increase in legibility and loudness.

The use of filters or a computer when processing the tapes can be a helpful tool to find the voices. See in the practical projects how to install filters and how to convert the recordings into computer files.

Imagination is important when performing the experiments. There are researchers who found the voices on the reverse side of the tape! You can move the noise source to find the best position in a room or even use more than one noise source.

1.10 Interpretation

Projection is a natural human tendency. By *projection*, we refer to the process by which something pertinent is created and "heard" in our minds that is not actually present on the tape. By this process, we might hear only a single syllable and project it into a complete word that is not present in fact. The brain attempts to transform the illogical into the logical.

Because of this, we must take extreme care when listening to the tapes. Not every voice or every word is going to make sense. Not every phrase will be understood, and often what it is heard has nothing to do with us directly. Keeping this is mind is very important to safeguard against tricking yourself. It is not out of line to recommend that the reader gain a basic understanding of the human perceptive functions as part of this research. And if you intend to write a scientific chronicle on this subject, remember to avoid beginning your experiments with preconceived ideas.

Another human natural tendency is not to explain the facts according to what they actually mean but to adapt the facts to explain the theories we hope to prove. In particular, this kind of error is found among many religious and mystic groups working with ITC.

1.11 Starting Up

The projects described in the following pages start from the simplest and move to the more complex. For readers who are inexperienced in electronics, we recommend beginning with the simplest or even to seek the assistance of a technician.

It is important that readers who want to install their own circuits have some experience with basic electronic projects, recognizing components, and knowing how to use the soldering iron. A place to work and some appropriate tools are also important.

The reader also must know how to read a schematic diagram and how to use printed circuit boards. Many books about this subject have been published that can help the reader who wants to fabricate custom circuits.

1.12 Using a Tape Recorder to Pick Up "The Voices"

The simplest way to pick up the voices is shown in Fig. 8. The natural background noise is used as a "carrier" to reveal the presence of undetectable signals in an ambient.

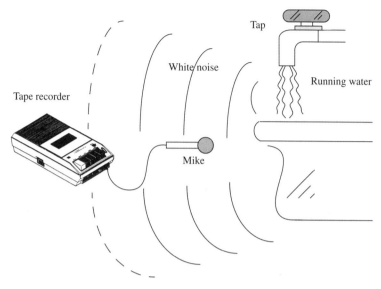

Figure 8 The simplest experiment.

Use the tap running in an ambient as the source of white noise. Other noise sources can be used, such as a fan or an AM radio tuned between stations. We have heard about an individual who conducted experiments during an hurricane! It is important to call the reader's attention to the fact that the voices are very weak, and trained ears are necessary for their identification in most cases (see section on "Performing the Experiments"). In many cases, only a few words will become clear among many minutes or hours of tape recorded.

Some researchers have upgraded their systems using a germanium diode plugged into the tape recorder's microphone input (1N34 or equivalent, for instance). The noise generated by this component, and also ambient detected radio signals, are used as "carrier" when picking up the voices.

Of course, more sophisticated circuits can be used if the reader has a basic knowledge in electronics. Another suggestion is to leave a fan running as a wind source, which also can be an excellent white noise generator for the experiments as shown in Fig. 9.

Suggestions

- Try using other noise sources, such as a resistor or even an antenna plugged into the input of the circuit as shown in Fig. 10.
- Natural noise sources such as wind and rain can also be used in these experiments.
- Avoid the use of long cables plugged to the input of the circuit, as hum (60 Hz signal from the ac power line) can be picked up, affecting the results.

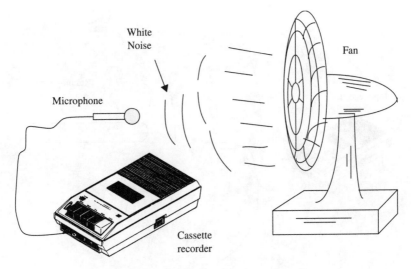

Figure 9 Another noise source suitable for these experiments.

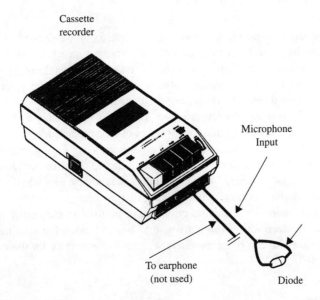

Figure 10 Using a diode as a noise source.

- Another noise source is your AM radio receiver. Tune it between stations and raise the volume. The "hiss" produced by the loudspeaker is a form of white noise. Setting old TV sets between channels or on free channels provides you with another white noise source. (New TV sets have a "squelch" circuit that cuts off the sound when no signal is picked up from a station.)

	Parts List: Using a Tape Recorder
1	Tape Recorder (cassette or reel-to-reel)
1	Microphone (included with the tape recorder or external, preferred)
1	Diode 1N34 or equivalent (optional)
1	Noise source (see text)

1.13 White Noise Generators

The simplest white noise generator is probably the one shown in Fig. 11. This circuit consists in a base-emitter junction of silicon transistor used as noise source, a power supply, a bias resistor, and a capacitor to couple the signals to an audio amplifier. The power supply is formed by a battery with voltages starting at 9 V. The best results are realized with 12 V or more.

In some cases, this circuit doesn't have power enough to drive common amplifiers, and amplification stages are needed. So, the following projects are recommended to easily drive common amplifiers.

Project 1: White Noise Generator I

The first important circuit to be suggested is the white noise generator shown in Fig. 12. This circuit will produce an electric signal without a definite frequency. When amplified and used with an earphone or loudspeaker (via an amplifier), the electric noise will produce an audible white noise. Note that white noise can be defined both as electric signals and mechanical vibrations in the audible spectrum.

How It Works

The thermal noise in a semiconductor junction (Q1) is amplified by the circuit and can be used as carrier or background noise source in the experiments. As a

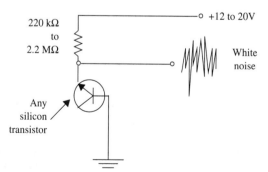

Figure 11 The simplest white noise generator.

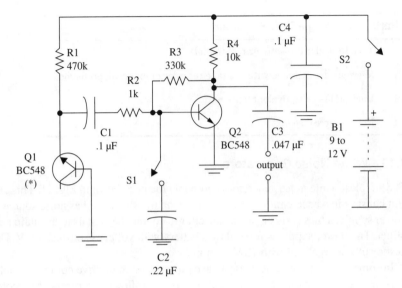

Figure 12 White noise generator I.

thermal source of white noise we can use any silicon diode such as the 1N4148 or 1N914, or the junction of any silicon general-purpose NPN transistor such as the 2N2222, BC547, BC548, 2N3904, etc.

Closing switch (S1) you can change the dynamic range of the noise, transforming the circuit into a "pink noise" generator. As mentioned earlier, the difference between the two kinds of noise is that white noise has a constant amplitude along its frequency spectrum, whereas pink noise has an amplitude that decreases with frequency.

Transistor Q2 is used to increase the amplitude of the noise acting as an amplifier. The gain is determined by R3 and R4.

Assembly

The circuit is not critical. You can mount it on a printed circuit board, a terminal strip, or even a solderless board. Figure 13 shows the layout of a printed circuit board that can be used to mount this circuit.

The only precaution you must take is with the output cable. You may use a shielded cable with a plug that matches the input jack of your tape recorder or amplifier. If long cables or improperly shielded cables are used, the circuit can pick up hum (60 Hz from the ac power line), affecting the results.

The power supply can be a 9 V cell or eight AA (12 V) cells. The best results are usually obtained with a 12 V power supply, depending on the transistor used as the noise source. You can use two four-cell battery holders wired in series to obtain a 12 V source.

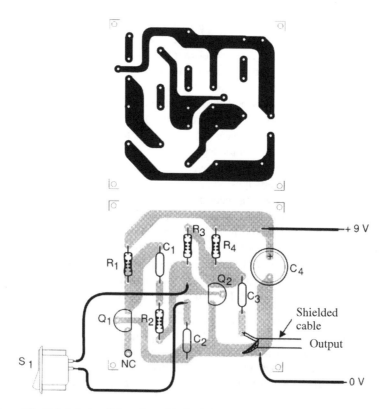

Figure 13 Printed circuit board used in Project 1.

Using the Circuit

The white noise generator can be plugged to the input of any audio amplifier as shown in Fig. 14. Another possibility is to plug the circuit to the input of a tape recorder, transferring the signals directly to the circuit.

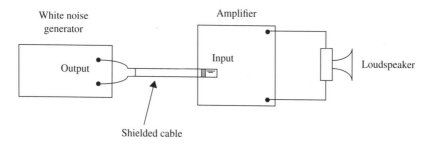

Figure 14 Using the circuit.

In the case of the tape recorder, record on virgin tape with the circuit plugged into the microphone input. Rewind the tape, and then try to hear some sounds mixed with the white noise. Remember that the sounds or voices are very weak and sporadic. It is necessary to be patient, and many hours of recording are necessary to find something interesting. Refer to the experimental procedures outlined at the beginning of this section. The use of a headphone can be interesting and helpful when trying to find some voices on the tape.

In the case of the amplifier, the circuit can be used to fill an ambient with white noise, reproduced by a common loudspeaker.

Although the suggested circuit is powered from a 9 V battery or 8 AA cells, the reader can also use a power supply such as the one shown in Fig. 15.

Any small transformer with a secondary winding rated to currents from 100 to 500 mA and voltages between 12 and 15 V can be used. Take care when using this circuit to avoid shock hazards, as it is plugged directly to the ac power line.

IC1 is not critical, as the circuit can be powered from a wide range of voltages. So, for IC1 you can use the 7812 or 7815. The suffix indicates the output voltage. For instance, the 7812 gives an output of 12 V. This component doesn't need a heatsink, as current drain is very low when powering this circuit.

Important: when using the power supply, the circuit becomes more sensitive to the "hum" of the ac power line. This hum is a 60 Hz signal that can cover up the white noise as well as the voices, compromising the experiments. First ensure that the shield is solidly connected to the ground of the circuit (0 V point).

There are several ways to reduce or eliminate the hum, as follows:

1. Keep the cable between the white noise generator and the amplifier or tape recorder short.

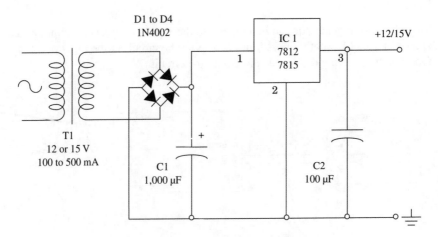

Figure 15 Power supply recommended for Project 1 and others.

2. Invert the position of the power supply plug (i.e., flip it 180° and plug it into the ac power line again).
3. Power the tape recorder from cells rather than the ac power line or use a higher quality power supply for the tape recorder.
4. Move the equipment to locations far from the sources of hum, which include the wiring of your home and household appliances such as refrigerators, microwave ovens, fans, etc.

You can also use a hum filter between the input circuit and the tape recorder or the amplifier. One of our projects is a noise filter that can be tuned to 50 or 60 Hz. (In some countries, the ac power line uses a 50 Hz current.)

Suggestions

- Change R1 within the range of 220 kΩ to 1.2 MΩ to get the better performance according to the semiconductor (Q1) used as white noise source.
- Vary C2 in the range between 1,200 pF and 0.1 µF to alter the dynamic range of the pink noise.
- R3 can be changed within a range between 330 kΩ and 1.2 MΩ to alter the gain of the transistor. Place in series with C2 a 100 kΩ potentiometer.
- Place in series with C2 a 100 kΩ potentiometer. You can use this potentiometer to control the dynamic range when the circuit is producing pink noise.

Parts List: Project 1

For the Circuit

Semiconductors

Q1	BC548, 2N3904, or any transistor, or even a silicon diode (see text)
Q2	BC548, 2N3904 or 2N2222, general-purpose NPN silicon transistor

Resistors

R1	470 kΩ, 1/8 W, 5%—yellow, violet, yellow (see text)
R2	1 kΩ, 1/8 W, 5%—brown, black, red
R3	330 kΩ, 1/8 W, 5%—orange, orange, yellow
R4	10 kΩ, 1/8 W, 5%—brown, black, orange

Capacitors

C1, C4	0.1 µF, ceramic or polyester
C2	0.022 µF, ceramic or polyester (see text)

C3	0.047 µF, ceramic or polyester

Miscellaneous

S1, S2	SPST, toggle or slide switch
B1	9 V battery, 8 AA cells, or power supply (see text)

Printed circuit board or solderless board, plastic box, battery clip, shielded cable, plug according to the input of the tape recorder or amplifier, wires, solder, etc.

Power Supply

IC1	7812 or 7815, voltage regulator, integrated circuit (see text)
D1 to D4	1N4002 or equivalent, silicon rectifier diodes

Capacitors

C1	1,000 µF, 16 or 25 WVdc, electrolytic
C2	100 µF, 16 WVdc, electrolytic

Miscellaneous

T1	12 to 15 V, 100 to 500 mA secondary, transformer

Plastic box, power cord, wires, solder, etc.

Project 2: White Noise Generator II

The level of white noise produced by the circuit shown as Project 1 is determined by the gain of the amplifier stage formed by Q1. In some applications, this noise level isn't enough to drive a low-sensitivity audio amplifier. To increase the output level of the circuit, we can add a second amplification stage such as the one described in this project. With this additional stage, the circuit can supply a stronger signal to the input of recorders, amplifiers, and transmitters.

How It Works

Q1 is the noise source or generator and is the same used in the previous project. You can use any silicon diode or silicon general-purpose NPN transistor.

Transistors Q2 and Q3 are the core of the amplification stage, increasing the noise amplitude to values up to half a volt, which is enough to drive external circuits such as amplifiers. As in the previous project, if S1 is closed, the circuit can be used as a pink noise generator.

The circuit can be powered from a 9 V battery or a power supply that provides voltages between 9 and 12 V, such as the one suggested for Project 1.

Depending on the transistor used as a noise source, the circuit may not operate with 9 V supplies and may require a 12 V supply. In this case, you can use eight A cells as in the previous project. The current drain is very low, extending the battery life if a battery is used.

Assembly

The circuit of the white (and pink) noise generator is shown in Fig. 16. Placement of the components on a small printed circuit board is shown in Fig. 17. An alternative mounting process is the one that uses a terminal strip as the chassis. The use of discrete parts such as transistors, resistors, capacitors, and other components (i.e., no IC) allows the reader to choose this solution for mounting. For experimental purposes, the reader can also mount the circuit on a solderless board.

The circuit is not critical, but special care must be taken with the output cable. It must be shielded to avoid hum (a 60 Hz signal that can be picked up from the power supply line). This care must be doubled if the circuit is mounted on solderless board or if the circuit is powered from the ac power line.

The circuit and power supply can be housed in a small plastic box. A metallic box is also suitable, with the advantage that it will act as an electromagnetic shield, cutting external noise and hum.

Using the Circuit

Plug the circuit to the microphone input of a tape recorder or an audio amplifier as in the previous project. In the first case, the noise is applied directly to the tape.

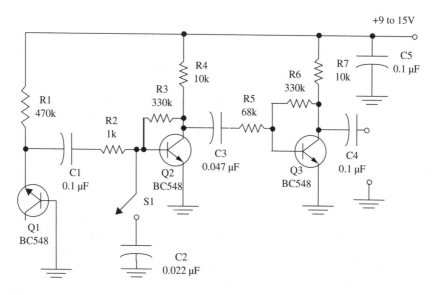

Figure 16 White noise generator II.

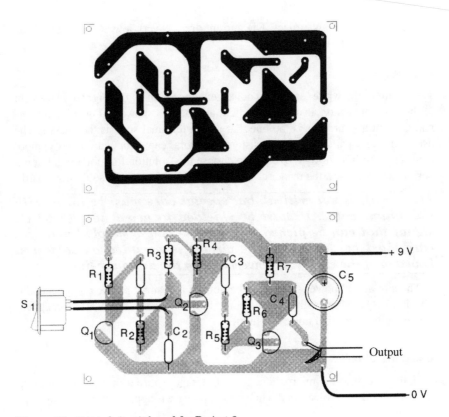

Figure 17 Printed circuit board for Project 2.

In the second case, the amplifier with a loudspeaker can be used to fill an ambient with white and pink noise. The circuit can also modulate transmitters, as will be described in the experiments involving radio waves.

The noise level is adjusted by the volume control in the amplifier. The circuit can be powered by battery, AA cells, or the same power supply recommended for the previous project.

Suggestions

- The same alterations suggested in Project 1 can be used in this project.
- R6 can also be varied between 330 kΩ and 1 MΩ to alter the circuit gain.
- A 10 kΩ potentiometer can be added to the output of the circuit as a volume or amplitude control. The wiring of this control is shown in Fig. 18. It is important when the signal is injected directly into the input of a tape recorder to avoid saturating (overdriving) the circuits.
- Try using a dynamic filter to the noise frequency spectrum as shown in Fig. 19.

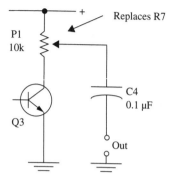

Figure 18 Adding a volume control.

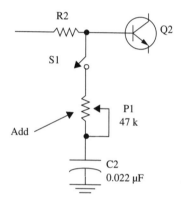

Figure 19 Adding a control to the filter.

Project 3: White Noise Generator Using an IC

The circuit shown in Fig. 20 produces a strong white noise signal covering a frequency band between a few hertz and more than 100 kHz. Adaptations can change the dynamic range for pink noise generation.

The operating principle is the same as with the previous circuits: a silicon diode or the junction base-emitter of a silicon transistor is used as the noise source.

In this circuit, the white noise signal is amplified by an operational amplifier IC 741 (NE741, LM741, uA741, or any equivalent). You can also try using JFET operational amplifiers such as the CA3140, TL070, TL080, or others. The negative feedback, and therefore the gain, is controlled by adjusting the potentiometer P1.

The operational amplifier has a voltage gain of up to 100,000×, giving an excellent output to drive amplifiers or other audio circuits. The output impedance is low (about 150 Ω), representing a high power gain.

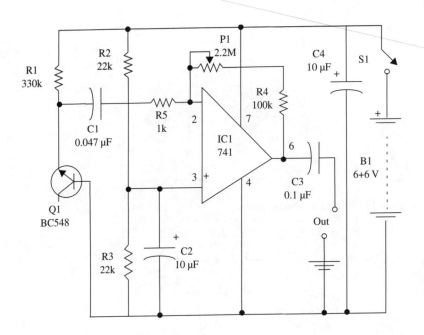

Figure 20 White noise generator with IC.

This circuit must be powered from a 12 V source. To eliminate the need of a symmetrical power supply, the bias is given by a voltage divider formed by R2 and R3. These components are not critical, and values between 10 kΩ and 100 kΩ are suitable for this task.

The power supply is formed by eight AA cells (12 V) or two 9 V batteries wired in series. However, a supply from the ac power line can be used. In this case, the voltage can range from 12 V to 18 V, according to the voltage regulator integrated circuits used in the project. The suffix of the ICs given in the parts list indicates the output voltage. For instance, the 7812 produces 12 V at its output.

As the current drain is very low at the output, the IC doesn't need a heatsink. The battery power supply is preferred to avoid hum.

Assembly

This circuit can be mounted on a small printed circuit board, including the power supply, or for experimental purposes on a solderless board.

All the components, including the power supply (cells or batteries), can be housed in a small plastic box. A metallic box is recommended, as it can act as a shield to avoid hum.

Any transformer with secondary voltages between 12 and 15 V is suitable for a power supply powered from the ac line.

Parts List: Project 2

Semiconductors

Q1	Any silicon general-purpose NPN transistor: BC548, 2N3904, 2N2222, etc. (see text)
Q2, Q3	BC548, 2N222, 2N3904, or any general-purpose silicon NPN transistor

Resistors

R1	470 kΩ, 1/8 W, 5%—yellow, violet, yellow
R2	1 kΩ, 1/8 W, 5%—brown, black, red
R3, R6	330 kΩ, 1/8 W, 5%—orange, orange, yellow
R4, R7	10 kΩ, 1/8 W, 5%—brown, black, orange
R5	68 kΩ, 1/8 W, 5%—blue, gray, orange

Capacitors

C1, C4, C5	0.1 µF, ceramic or polyester
C2	0.22 µF, ceramic or polyester (see Project 2 for details)
C3	0.047 µF, ceramic or polyester

Miscellaneous

S1, S2	SPST, toggle or slide switch
B1	9V battery, 8 AA cells (12V) or power supply (see text)

Printed circuit board, shielded cable and jack plug according to the amplifier or tape recorder, plastic box, battery clip, 2 × 4 AA cells holder, solder, wires, etc.

Care must be taken with the output cable. You must use a shielded cable to avoid hum (see Projects 1 and 2 for more details).

Using the Circuit

This circuit is used as in the previous projects. The white noise is applied to the input of tape recorders, amplifiers, or other circuits and used according to the experiments to be performed. P1 is adjusted to determine the noise amplitude of the produced signal.

Suggestions

- Vary the value of R1 in the range between 220 kΩ and 2.2 MΩ to increase the noise level.
- R3 and R4 can be varied in the range between 10 kΩ and 100 Ω.
- Use a wire between pin 2 and 6 in the network shown in Fig. 21 to add a pink noise option to the circuit. The capacitor can have values between 1,000 pF and 0.47 µF. Closing S1, the circuit will generate pink noise.

1.14 Experimenting with White Noise Generators

Projects 1 through 3 are white and pink noise generators recommended for use in several experiments with the aid of amplifiers, tape recorders, filters, transmitters, computers, graphic equalizers, or other equipment.

Some experiments can be performed in addition to the ones described previously.

- Basic configuration: using an audio amplifier (any of your sound equipment), you can fill an ambient with white noise or pink noise and let your tape recorder record with a microphone to pick the voices. Figure 22 shows how the experiments could be performed. Make sure that the white/pink noise generator is plugged to the auxiliary input (AUX) and that the noise level is adjusted by the amplifier and by the generator (if available).
- If you are using a stereo amplifier, you can use two noise sources as shown in Fig. 23, plugging them to both channel inputs (left and right), or use one input using the amplifier in the MONO option.
- A germanium diode such as the 1N34 can be added to produce a special effect, rectifying the output of the white/pink noise generators signal as shown in Fig. 24.

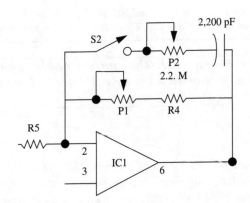

Figure 21 Adding a bandpass control to the circuit.

Parts List: Project 3

Semiconductors

Q1   Any silicon NPN transistor or silicon diode (BC548, 2N3904, 2N22222) (See Project 2).

IC1   741 or any equivalent operational amplifier integrated circuit

Resistors

R1   330 kΩ, 1/8 W, 5%—orange, orange, yellow

R2, R3   22 kΩ, 1/8 W, 5%—red, red, orange

R4   100 kΩ, 1/8 W, 5%—brown, black, yellow

R5   1 kΩ, 1/8 W, 5%—brown, black, red

Capacitors

C1   0.047 µF, ceramic or metal film

C2, C4   10 µF, electrolytic

C3   0.1 µF, ceramic or metal film

Miscellaneous

P1   2.2 MΩ potentiometer

B1, B2   6 or 9V, AA cells, battery, or power supply (see text)

S1   SPST, toggle or slide switch

Printed circuit board or solderless board, shielded cable, wires, etc.

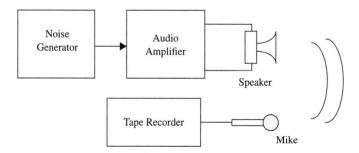

Figure 22   Basic configuration.

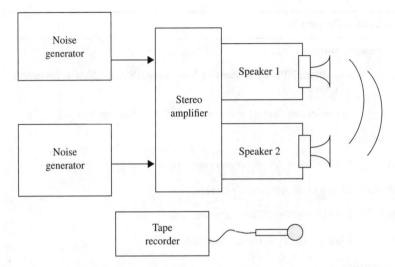

Figure 23 Using two noise sources and a stereo amplifier.

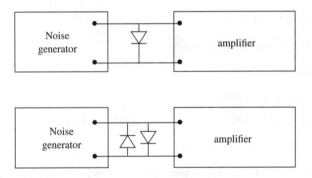

Figure 24 Using diodes to clip the noise.

Project 4: Sound Filter

Perfect sound reproduction cannot come from a tape, loudspeaker, or earphones. The output power is low, and the quality of the device (earphone or loudspeaker) may not be high enough to make it easy to find the voices. Separating the voices from the noise isn't a simple task, and it requires much concentration and patience. However, you can increase the performance of your circuit when reproducing the tapes by adding an audio output filter.

The first circuit of an audio filter shown here is very simple and uses only four components. But the main attraction in this project is that it doesn't need any power supply, because it is a passive filter.

How It Works

As the human voice has a frequency spectrum concentrated in the medium frequency band as shown in Fig. 25, when trying to detect the voices, the best results are obtained when high and low frequencies are cut off.

The passive filter circuit shown in Fig. 26 is plugged between the output of a tape recorder and the earphone or the input of an amplifier (such as those shown for earlier projects). It is a *bandpass* filter, which is a circuit that lets signals pass within a specified frequency range but cuts them off above and below.

The low-frequency signals are cut by an amount determined by the adjustment of P1, and the high-frequency components of the signal are cut according to the value of L1. By adjusting the circuit, you can let pass only the frequencies where the voice signals are concentrated, making it easier to find them.

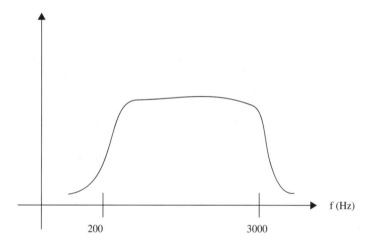

Figure 25 Human voice spectrum.

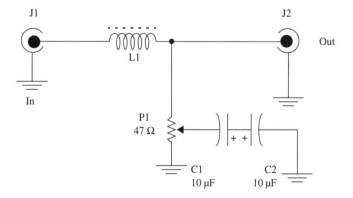

Figure 26 Sound filter.

Assembly

L1 is formed by 200 to 500 turns of wire, between 28 and 32 AWG, in a ferrite rod of 5 to 10 cm length (any diameter between 0.8 and 1.2 cm is suitable).

All the components can be housed in a small plastic box. To plug the circuit into the output of the tape recorder (monitor), use an appropriate plug and a shielded cable with a length up to 50 cm. To plug a headphone into the circuit, use a jack appropriate for the headphone plug. If you intend to plug it into the input of an amplifier, use a shielded cable with appropriate plugs.

The correct use of this circuit is between the output of the amplifier or tape recorder and a low-impedance transducer such as a loudspeaker or an earphone.

When installing the filter between a low-impedance source and a high-impedance device such as an amplifier, distortion due to impedance mismatch can occur. In this case, wire a 100 Ω × 1 W resistor in parallel with the filter's output.

Using the Circuit

Plug the filter into the output of the tape recorder and let the machine record in the location where you intend to find the voices. Plug the earphone into the output of the filter. Note that this circuit is designed to be compatible with the tape recorder and is designed for low-impedance earphones of 8 to 100 Ω. If you're using an audio amplifier, plug the output of the filter into the AUX (auxiliary) input of the amplifier.

Set the tape to run and adjust the filter control, trying different positions while looking for the voices. If you are using an audio amplifier, adjust its volume to obtain the best results.

Suggestions

- The number of turns of L1 can be altered to change the filter performance. Make such experiments, or use different filters (with different coils).
- Capacitors C1 and C2 determine the high-frequency cutoff. Change them, experimenting with values between 1 and 100 µF.

Parts List: Project 4

L1	Coil (see text)
C1/C2	10 µF/16 WVdc, electrolytic capacitors
P1	47 Ω, wire wound potentiometer
J1, J2	Input and output jacks or plugs according to the tape recorder, earphones, and amplifier (see text)

Ferrite rod, plastic box, knob for P1, wires, solder, etc.

Project 5: Noise Filter

This circuit can be used to eliminate the background noise on a tape, revealing voices and other sounds more easily. It will be installed between the tape recorder and an amplifier as shown in Fig. 27. The basic idea is the use of a noise source to add energy to the undetectable voices (see stochastic resonance), increasing their level when taped. Then, when listening to the tape, the researcher can reduce or eliminate the background noise to more easily find the voices.

Operation

The circuit is an amplifier stage that cuts the signals between two silent zones or between two detectable signals. This means that this circuit cuts the signal when the ratio of signal to noise falls bellow a certain level.

With the values shown in the diagram of Fig. 28, the circuit cuts signals with amplitudes below 30 mV. If a signal with an amplitude greater than 30 mV is

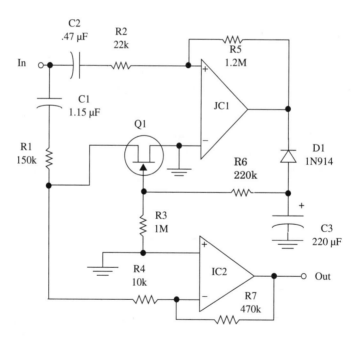

Figure 27 Noise filter.

Figure 28 Operation of the noise filter.

present, operational amplifier A1 has a high enough output to bias Q1. This implies an increase in gain at the operational amplifier A2, and the amplified signals appear in its output.

When the amplitude of the input signal is below 30 mV, Q1 is not biased and the gain of the circuit is reduced to less than 1/40 of the initial value.

To operate with signals over 30 mV, it is enough to increase the value of R2. Values between 10 kΩ and 100 kΩ can be used experimentally.

The gain of the circuit is determined by R7. Values between 100 kΩ and 1 MΩ can be used experimentally.

The circuit needs a symmetric, or two-voltage, positive/negative power supply as shown in Fig. 29.

Voltages between 9 and 15 Vdc can be used to power this circuit. Current drain is very low, less than 10 mA, allowing the use of batteries as power supplies. Some operational amplifiers can be powered from lower voltages. The CA3140, for instance, can be powered from voltages as low as 3 V. IC1 and IC2 don't need a heatsink, as the current drain is very low.

Assembly

The circuit can be mounted on a small printed circuit board and housed in a plastic box. Input and output jacks should be installed to match the output of the tape recorder and the input of the amplifier. The audio signals must be transferred to the circuit and from the circuit using shielded cables.

Any JFET operational amplifier such as the CA3140, TL071 or the dual TL072 can be used. The transistor specifications also allow equivalents such as the MPF102 or any JFET transistor. Note that the terminal placement of the MPF102 is different from that of the BF245.

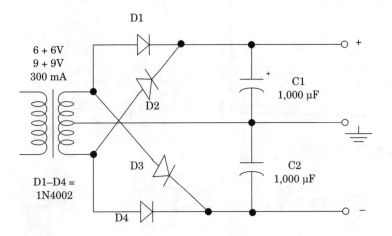

Figure 29 Symmetric or two-voltage source for operational amplifiers.

As the circuit has a high-impedance input and high gain, special care with the signal cables is needed to avoid hum. As in the other cases, the hum (60 Hz noise picked up from the ac power line) can cover the noise as well as the voices recorded on the tape, compromising the experiments.

Using the Circuit

The circuit is placed between the signal source (tape recorder) and the amplifier as shown in Fig. 28. The circuit will cut noise, allowing some sound peaks through. If no sound is heard, you must increase the output level of your tape recording, trying to find a level that will give the best voice detection results. Careful experimentation must be conducted to find the ideal point.

Notice that this circuit has a high-impedance output and input. This means that earphones, headphones, and loudspeakers can't be driven directly from its output. If the circuit tends to distort when plugged into the output of a tape recorder (monitor) add a 100 Ω resistor in parallel with the filter's input.

Initial operation tests can be made using a tape with common content such as music or voice. Voice is recommended, as the intervals between words will engage the circuit, cutting the background noise.

Suggestions

- Alter the value of resistors R2 and R7 to change the performance to get the desired effects.
- C3 determines the frequency characteristics of the circuit, and values between 100 µF and 1,000 µF can be used experimentally.
- Replace R7 with a 1 MΩ potentiometer, which can be used as a volume control.
- R5 can also be varied using resistors rated between 100 kΩ and 1 MΩ to alter the behavior of the noise detection.

Parts List: Project 5

Semiconductors

IC–1	CA3140, TL071, or TL072; JFET operational amplifier	
Q1	MPF102, BF245, or equivalent; any JFET transistor	
D1	1N914, 1N4148, etc., general-purpose silicon diode	

Resistors

R1	150 kΩ, 1/8 W, 5%—brown, green, yellow	
R2	22 kΩ, 1/8 W, 5%—red, red, orange	
R3	1 MΩ, 1/8 W, 5%—brown, black, green	

R4 10 kΩ, 1/8 W, 5%—brown, black, orange

R5 1.2 MΩ 1/8 W, 5%—brown, red, green

R6 220 kΩ, 1/8 W, 5%—red, red, yellow

R7 470 kΩ, 1/8 W, 5%—yellow, violet, yellow

Capacitors

C1 0.15 µF, ceramic or metal film

C2 0.47 µF, ceramic or metal film

C3 220 µF, 25 WVdc, electrolytic

Miscellaneous

Printed circuit board, plastic box, input and output jacks, power supply, wires, shielded cable, solder, etc.

Project 6: 60 Hz Filter (Hum Filter)

The "hum" or 60 Hz noise is one of the most terrible of the enemies of all audio equipment. The wires of our home power system act as antennas, irradiating a 60 Hz signal from the ac power line in all directions. All sensitive circuits, such as audio amplifiers, radios, and even your telephone line, can pick up these signals and amplify them. The result is their detection and reproduction in a loudspeaker, headphone, or the telephone as a continuous "hum." (See Fig. 30.)

If you touch a bare wire that is plugged to the input of any audio amplifier, you can hear this hum. The sound comes from the 60 Hz signal picked up by your

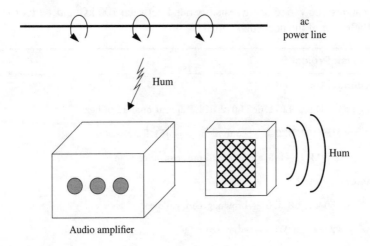

Figure 30 Sensitive audio equipment can pick up the ac power line signal.

body and amplified by the circuit. To avoid the hum, the circuits must be shielded. But, in some cases, even this procedure is not enough to block the incoming 60 Hz signals.

The hum can enter via an improper filtering circuit of a power supply, low-signal circuits, or even when you use a long cable to plug a microphone into the circuit or to transfer signals from one piece of equipment to another. A few microvolts of difference in the ground potential levels of two devices is enough to induce hum. For the ITC researcher or those who are trying to find the best combination of audio equipment for making these experiments, the noise can become a serious problem. As Fig. 31 shows, two devices can be connected to a common ground to avoid hum.

How It Works

The circuit is formed by an FET amplifier stage with strong negative feedback made by a resonant twin-T circuit. The feedback circuit feeds the signal back to the output only if its frequency is 60 Hz. This means that the gain of the circuit is reduced for 60 Hz signals but kept normal with other signals.

The potentiometer is used to tune the circuit to the exact frequency to be rejected—60 Hz, or 50 Hz if this is the frequency of your ac power line.

Notice that this circuit has a high-impedance input and can't be used directly with transducers such as loudspeakers or earphones.

The circuit also provides a gain to the input signal. Therefore, it can be used also as a preamplifier for microphones when installed between the amplifier and the microphone or other audio sources.

Assembly

The circuit of the Hum Filter is shown in Fig. 32. The circuit can be mounted on a small printed circuit board and installed in a metallic or plastic box. Metallic boxes are preferred, as they can act as a shield to prevent the circuit from picking up external noise. Jacks are used to plug the circuit to the audio source and to the input of an amplifier.

Any NPN general-purpose JFET transistor can be used in the project. Types such as the BF245, MPF102, or any other can be used. Note that the transistor,

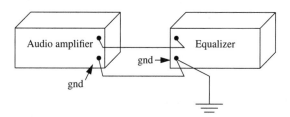

Figure 31 A common ground provides a solution to hum problems in audio systems.

40   Part 1

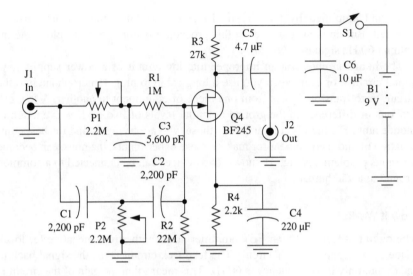

Figure 32 Hum filter.

depending on the type, may have different terminal placements. The patterns shown in the figure are for the BF245.

The circuit is powered from a 9 V battery. As the current drain is very low, the battery life can be extended to several weeks. Keep the wires short to avoid hum after the amplification stage!

Using the Circuit

Plug the output of the tape recorder into the input of the filter if the tape is found to have a high level of hum. Other audio sources with high hum levels can be plugged into the input, including sensors, microphones, etc.

Adjust P1 to block the hum. If you can not find the correct setting when adjusting these components, alter the values of C1, C2, and C3. Values 50% higher or lower than the suggested ones can be used experimentally. The values of the capacitors must maintain the same ratio as the original values. This means that if you multiply by 2 the value of C1, you must also double the values of C2 and C3 when performing these experiments. We must remember that many electronic components, including capacitors, have high tolerances for which we must compensate in some cases.

Suggestions

- This circuit can be placed between a microphone and a tape recorder input if you are using long cables and hum is picked up. For many applications, such as when recording with a long microphone cable, this filter can be quite helpful.

- Replace R8 with resistors rated between 15 and 47 kΩ if distortion appears in the sounds.
- Use other values for C3, C4 (0.1 µF), and C2 (0.22 µF) if you intend to use the device in a 50 Hz power line.

Parts List: Project 6

Semiconductors

Q1	BF245 or equivalent, JFET

Resistors

R1	1 MΩ, 1/8 W, 5%—brown, black, green
R2	22 MΩ, 1/8 W, 5%—red, red, green
R3	27 kΩ, 1/8 W, 5%—red, violet, orange
R4	2.2 kΩ, 1/8 W, 5%—red, red, red

Capacitors

C1, C2	2,200 pF, ceramic or metal film
C3	5,600 pF, ceramic or metal film
C4	220 µF/12 WVdc, electrolytic
C5	4.7 µF, 12 WVdc, electrolytic
C6	10 µF, 12 WVdc, electrolytic

Miscellaneous

P1	2.2 MΩ potentiometer
S1	SPST, toggle or slide switch
B1	9 V battery
J1, J2	Jack (see text)

Printed circuit board, plastic box, battery clip, wires, shielded cable, solder, plastic knobs, etc.

Project 7: Low-Impedance Preamplifier

Some transducers used in ITC experiments are low-impedance devices that are not powerful enough to drive amplifiers, tape recorders, or other equipment properly when directly plugged to them. Such transducers require the assistance of a low-impedance preamplifier to operate with such equipment.

The preamplifier must be installed between low-impedance transducers such as microphones, pickup coils, sensors, and other signal sources and the medium/high-impedance input of an audio amplifier or other equipment. The circuit can be used with transducers or signal sources with impedances ranging from 2 to 500 Ω.

The power supply is a 9 V battery, and power consumption is very low. We don't recommend a power supply plugged to the ac power line, as hum can be introduced if the output is not well filtered.

How it Works

The circuit is formed by a single common-base stage with an NPN high-gain transistor. The low-impedance audio source is plugged into the emitter of the transistor where the signal is applied. The amplified signal appears in the collector from where it can be transferred to the output via C3. Bias is given by R1 and decoupling by C1. R3 fixes the input impedance and also the gain. This resistor can be replaced by others in the range between 47 and 1 kΩ, depending on the impedance of the signal source. The reader must experiment with different values of R3 to get better performance, which depends on the signal source.

Assembly

The circuit is shown in Fig. 33. Keep the wires short, and use a shielded cable for the output signal. The input and output jacks are chosen to match the signal source and the input of the amplifier or tape recorder. The unit can be housed in a small plastic box with the battery. The components are not critical, and equivalents of the indicated transistor can be used for experimentation.

Using the Circuit

The circuit is placed between the audio source and the input of the amplifier or tape recorder. The output must be plugged to the AUX or other high- or medium-impedance input of the amplifier.

For example, a pickup coil can be used if you want to try to intercept voices carried over a telephone line. Place the pickup coil near a common telephone device, adjusting for a position that gives the best results. The hum filter previously described can be useful in this experiment to block any 60 Hz noise that may be present in the line.

Suggestions

- A pickup coil can be used as the transducer to experiment with noises occurring in the ambient magnetic field. This is shown in Fig. 34. The coil is formed

by 1,000 to 10,000 turns of 32 to 34 AWG wire wrapped around a ferrite rod. You can also use the primary of a small transformer as a pickup coil. Remove the core and place a ferrite rod inside it.
- Small loudspeakers and low-impedance earphones can be used as microphones with this preamplifier.

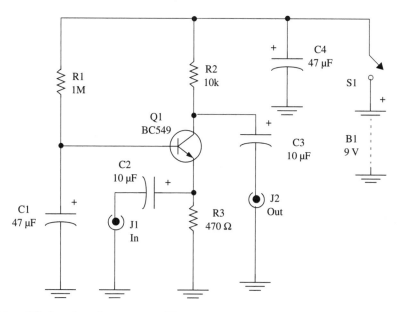

Figure 33 Low-impedance preamplifier.

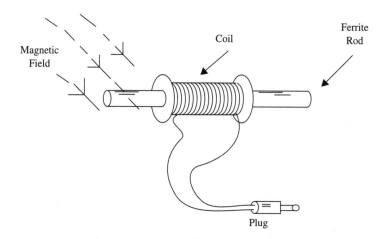

Figure 34 Picking up ambient magnetic fields.

Parts List: Project 7

Semiconductors

Q1   BF549, general-purpose silicon NPN transistor—low noise

Resistors

R1   1 MΩ, 1/8 W, 5%—brown, black, green

R2   10 kΩ, 1/8 W, 5%—brown, black, orange

R3   470 Ω, 1/8 W, 5%—yellow, violet, red (see text)

Capacitors

C1   47 µF, 12 WVdc, electrolytic

C2, C3   10 µF, 12 WVdc, electrolytic

C4   47 µF, 12 WVdc, electrolytic

Miscellaneous

J1, J2   Input and output jacks

S1   SPST, toggle or slide switch

B1   9 V battery

Printed circuit board, battery clip, plastic box, shielded cable, wires, solder, etc.

1.15 Ultrasonic Sources

White and pink are not the only noise types that can be used by the EVP researcher as a support or carrier in these experiments. The stochastic resonance effect needs some kind of energy to fill the ambient to manifest undetectable signals, carrying them to a higher level, as described already, but not necessarily in the audio frequency range.

Many experimenters use forms of energy other than audible sound waves. One of them is *ultrasonic* waves. Sounds at frequencies higher than the human sensory limit (that is, above approximately 18,000 Hz) can be used to fill an ambient and perform many experiments.

Some simple circuits can be used to fill an ambient with ultrasonics. Notice that these circuits produce mechanical (compression) waves in air in the range between 18,000 and 30,000 Hz rather than magnetic fields or electromagnetic waves in the same frequency range.

Beat

Beat is a phenomenon that takes place when vibrations at two different frequencies are combined at a certain point in space. At the meeting point, the two waves will act in a manner that causes them to produce two new vibrations. That is, the point will vibrate at a frequency that is the sum of the original frequencies and, at the same time, at a frequency that is the difference between the two original frequencies.

Notice that, if two ultrasonic frequencies are combined (e.g., 20,000 and 25,000 Hz), the sum (45,000 Hz) is higher than the audible limit of the human ear, but the difference (5,000 Hz) can be heard, because it is in the audible frequency range. When we fill an ambient with ultrasonics, the beat, along with other inaudible vibrations as produced by the EVP, can be changed to the audible range and registered on tape. Therefore, the use of ultrasonic sounds can extend the range of frequencies to be investigated by researchers and add new possibilities for experimentation.

Project 8: Low-Power Ultrasonic Source

The simple oscillator described here can fill small ambients with ultrasonic waves in the range between 18,000 and 30,000 Hz (depending on the transducer). Powered from common cells or a small power supply, the circuit is very compact and uses easy-to-find parts.

The researcher can place the circuit near the microphone or a tape recorder as shown in Fig. 35, adding a new form of energy to the experiments. This energy can be modulated by the voices and also combined by *beat* with the white noise needed for the experiments.

The experimenter can try several combinations of white/pink noise generators and the ultrasonic generator to make these experiments. Two ultrasonic generators, operating at different frequencies, can be used to create beats for experiments involving EVP.

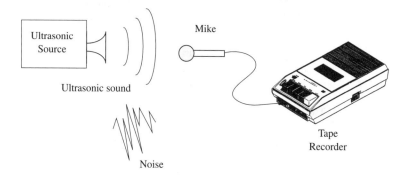

Figure 35 Using the ultrasonic sound source.

How It Works

The ultrasonic signal source is an oscillator based on one of the four Schmitt NAND gates of a CMOS 4093 IC. The frequency can be adjusted within a wide range of values by means of potentiometer P1.

With the values shown in the diagram, the circuit can produce ultrasounds in a range between 10,000 and 30,000 Hz. The audible part of the band is important, as it can be used to test circuit operation. The signals are amplified by the other three NAND gates wired as digital amplifiers.

For the transducer, we recommend the use of a piezoelectric tweeter. Many piezoelectric tweeters provide good performance when reproducing sounds between 18,000 and 22,000 Hz. However, to use the tweeter, some modifications must be made to this component.

The tweeter (high-frequency loudspeaker) is a low-impedance device, due to the presence of a small transformer inside. As our circuit has a high-impedance output, it is necessary to remove the small transformer inside the tweeter as shown in Fig. 36. Accessing the piezoelectric ceramic transducer, we can wire it directly to the output of our circuit.

The circuit can provide a few milliwatts of ultrasonics when powered from AA cells or a 9 V battery. This power is enough to conduct experiments in which it is placed near the microphone.

Assembly

The complete circuit of the low-power ultrasonic source is shown in Fig. 37. A small printed circuit board is used to mount the components.

The layout for this printed circuit board is suggested in Fig. 38. The components, including the piezoelectric tweeter and the battery holder, can be housed in a small plastic box.

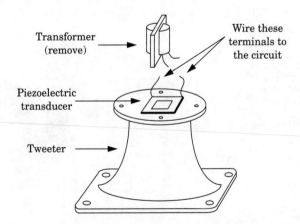

Figure 36 Adapting a piezoelectric tweeter for this application.

Instrumental Transcommunication 47

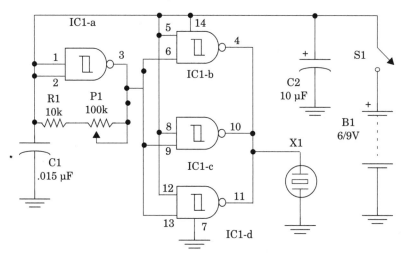

Figure 37 Ultrasonic source.

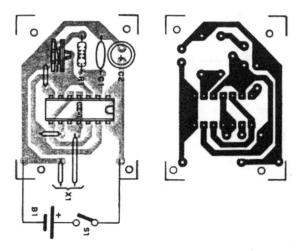

Figure 38 Printed circuit board for Project 8.

In the front panel we place the potentiometer where the operation frequency can be adjusted.

Using the Circuit

Turn on the circuit and adjust P1 until the audible sound increases in frequency and then disappears. This indicates that ultrasonics are being produced. Then place the ultrasonics source near the microphone used to pick up the voices (a distance between 30 and 80 cm is suitable for the experiments).

In the following pages, we will suggest some combinations of devices that include this oscillator for further experiments.

Suggestions

- Try using various small piezoelectric transducers as found in toys, alarms, and other applications. Some of them can provide reasonable performance for reproducing ultrasonics.
- Mount more than one unit of this circuit to try the beat effect, combining different frequencies when conducting the experiments. Each device will be adjusted by the corresponding potentiometer.
- The high-frequency signals produced by this circuit can be directed into the tape recorder or amplifier inputs using a mixer.

Parts List: Project 8

Semiconductors

IC1   4093, CMOS integrated circuit

Resistor

R1   10 kΩ, 1/8 W, 5%—brown, black, orange

Capacitors

C1   0.01 µF, ceramic or metal film

C2   10 µF, 12 WVdc, electrolytic

Miscellaneous

P1   100 kΩ, potentiometer

X1   Pieozelectric tweeter (without transformer) (see text)

S1   SPST, toggle or slide switch

B1   6 to 9 V, 4 AA cells or 9 V battery

Printed circuit board, battery clip or holder, plastic box, knob for P1, wires, solder, etc.

Project 9: High-Power Modulated Ultrasonic Source

The next circuit can produce several watts of ultrasonic waves, filling medium- and large-size ambients for experiments with the EVP. An additional high-power stage and a compatible power supply are added to the previous project. The circuit drains more current due to this power stage, so a power supply is needed. The circuit also includes a modulation stage. This stage, when activated, will turn the ultrasonic source on and off at regular intervals. Some experiments can be programmed using this effect.

As the output stage has a low impedance, it is not necessary to remove the transformer from the tweeter. Therefore, the transducer can be plugged directly to the output of the circuit.

Ultrasonics are dangerous to small mammals, as many of them can hear sounds up to 40,000 Hz (dogs, cats, rats, mice, etc.). A high-power ultrasonic source can cause severe discomfort in these animals, so avoid the use of this oscillator in places where they reside. Note: This circuit can be used to scare rats, mice, and other mammals from trash areas and other places where they congregate.

Operation

The circuit operates in the same manner as the previous one. We have only added a power output stage using a power FET, plus a low-frequency oscillator. The use of a power MOSFET adds to the circuit the capability of driving a low-impedance load from the signal found in the output of the digital amplifier using the 4093.

The additional stage is a low-frequency one, using one of the four NAND gates of a 4093 IC. P1 adjusts the modulation rate and P2 the ultrasonic frequency.

Assembly

The complete circuit of the high-power ultrasonic source is shown in Fig. 39. The components are mounted on a small printed circuit board as shown in Fig. 40.

Any power FET can be used in this circuit. Types with drain currents above 2 A and voltages above 100 V are suitable.

Q1 must be mounted on a heatsink. The heatsink can be simply a piece of metal bent to form a "U" or any commercial type as shown in the figure and affixed by a screw to hold the transistor in place.

A plastic box can be used to house the components. The tweeter can be placed outside the box. A cable with plugs can be used to plug the circuit to the power supply. A suitable power supply for this project is shown in Fig. 41.

The transformer has a secondary winding with 12 to 15 V × 1 A, and the diodes are 1N4002 or equivalents.

Using the Circuit

The circuit is used in the same manner as the previous one. See the suggestions for experiments at the end of this chapter. If Q1 tends to heat excessively, wire a 10 Ω × 5 W resistor in series with the tweeter.

50 Part 1

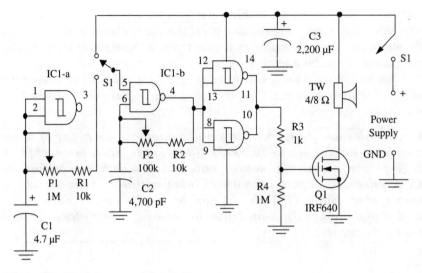

Figure 39 High-power modulated ultrasonic source.

Parts List: Project 9

Semiconductors

IC1	4093, CMOS integrated circuit
Q1	IRF640 or any equivalent (see text)

Resistors

R1, R2	10 kΩ, 1/8 W, 5%—brown, black, orange
R3	1 kΩ, 1/8 W, 5%—brown, black, red
R4	1 kΩ, 1/8 W, 5%—brown, black, green

Capacitors

C1	4.7 µF/12 WVdc electrolytic
C2	4,700 pF, ceramic
C3	2,200 µF, 25 WVdc, electrolytic

Power Supply

D1, D2	1N4002, silicon rectifier diodes
T1	Transformer, primary according the ac power line; secondary 6 or 9 V × 1 A

C1	1,000 µF/16 WVdc, electrolytic

Miscellaneous

P1	1 MΩ, potentiometer
P2	100 kΩ, potentiometer
TW	4/8 Ω, piezoelectric tweeter
S1	DPST, toggle or slide switch
S2	SPST, toggle or slide switch

Printed circuit board, heatsinks, fuse holder, power cord, wires, solder, plastic or metallic box, plastic knob, etc.

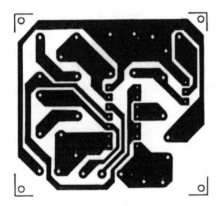

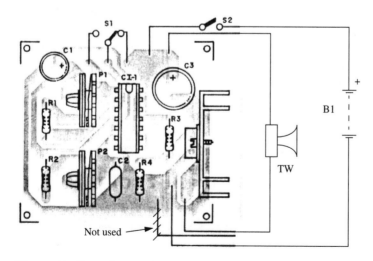

Figure 40 Printed circuit board used in Project 9.

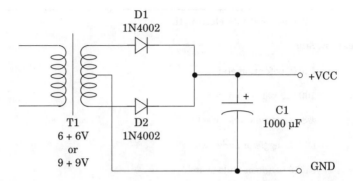

Figure 41 Power supply for Project 9.

1.16 Experiments

Detecting the voices is a difficult task, requiring a great dose of patience and a large number of experiments. Using the circuits described above and a tape recorder, the reader can perform many experiments such as suggested below.

1.16.1 Recording

Configuration 1

Figure 42 shows how a filter can be added to the input of the circuit. You can also place an equalizer (commercial) at point A. Experiments using several kinds of filters can be performed.

Configuration 2

Figure 43 shows how a mixer can be used to mix signals from the microphone with the signals coming from the noise generator.

In this case, the noise isn't picked up by the microphone but injected into the tape recorder input with the ambient sound. The researcher can also use an internal noise source simultaneously with an external noise source filling the ambient with white noise. A hum filter or an equalizer can be placed between the mixer and the tape recorder (point A).

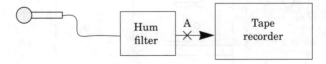

Figure 42 Configuration 1.

Instrumental Transcommunication

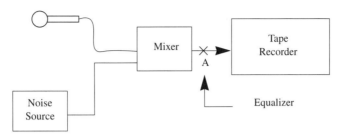

Figure 43 Configuration 2.

Configuration 3

Experiments using two different noise sources can be performed with the configuration shown in Fig. 44. You can experiment with several kinds of noise sources and pick up the signals using a microphone. The mixer can also be used with two microphones placed at different locations in a room.

Configuration 4

Figure 45 shows how two signal sources can be used in an experiment. In addition to the noise source (white or pink), we have an ultrasonic source. Both signals fill the ambient and are picked up by the microphone.

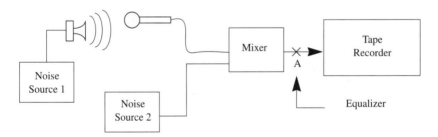

Figure 44 Configuration 3.

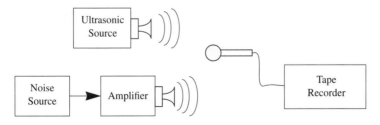

Figure 45 Configuration 4.

Configuration 5

Figure 46 shows how a pickup coil can be used to detect signals that are present in the ambient as low-frequency magnetic fields. The coil is formed with 1,000 to 10,000 turns of 32 to 34 AWG enameled wire in a ferrite rod. The preamplifier is the low-impedance preamplifier described earlier in this chapter. The coil must be wired to the circuit using a shielded cable.

To the noise picked up by the coil, some researchers add ambient radio signals from radio stations, detecting them with a diode. This diode (1N34, 1N60, or any other germanium type) is wired in series with the coil.

Configuration 6

Another way to fill the ambient with noise, but in the form of a low-frequency magnetic field, is shown in Fig. 47. The noise is amplified and applied to a coil that produces a low-frequency magnetic field. Placing the coil near another coil to pick up the signals, the noise can be mixed with any signal (picked up by a microphone or another signal picked up by the receiver coil).

The transmitter coil is formed by 50 to 200 turns of 26 to 28 AWG wire on a ferrite rod. The amplifier can be either the one using the LM386 or the TDA2002.

1.16.2 Processing

Configuration 7

Figure 48 shows the simplest configuration for processing the picked-up signals using a filter. Any of the previously described filters, and others, can be used. It is

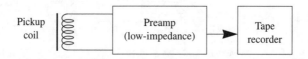

Figure 46 Configuration 5.

Figure 47 Configuration 6.

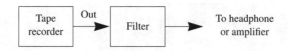

Figure 48 Configuration 7.

important to match impedances to get best results. The earphone must be a low-impedance type that matches the output of the tape recorder, normally between 8 and 100 Ω.

Configuration 8

Tape recorders usually have their own internal speakers, but we can't access them to place any kind of filter between them and the output circuit. In this case, we can use the tone control to help us find the voices, but tone controls are not very efficient. The use of an external audio amplifier makes it possible to place a filter between the audio source and the loudspeaker.

Any low-impedance filter, such as the ones described previously, can be used between the output of a tape recorder (earphone output) and the amplifier. In some cases, it will be necessary to place a 100 Ω resistor in parallel with the filter's output to avoid distortion.

Configuration 9

If hum appears in a tape, a hum filter can be placed between the tape recorder and the amplifier. In some cases, a 100 Ω resistor must be wired in parallel with the amplifier's input to avoid distortion.

Configuration 10

When analyzing a tape, a mixer can be useful to add signals to the picked-up sounds. The configuration shown in Fig. 49 shows how a noise generator can be plugged into an amplifier, adding its signal to a tape recorded sound.

1.16.3 Conclusion

The configurations shown are only some of the hundreds the reader can create using other devices such as commercial equalizers and mixers. The configurations used to make experiments are limited only by the reader's imagination. The cir-

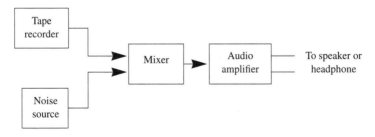

Figure 49 Configuration 10.

cuits shown previously are very flexible and noncritical. The reader can alter those circuits and try other combinations. Remember that the secret of success in this kind of research is found in the correct combination of circuits that makes it possible to tune in the voices.

1.17 Picking Up Sounds from the Earth

Our planet acts as a great conductor in which current induced by natural electric discharges, man-made equipment, and unknown sources flows in undetermined directions. By inserting two electrodes at various locations, as shown in Fig. 50, and plugging them into the input of an audio amplifier, we can hear low-frequency currents in the form of sounds.

Mixed with the background noise, it is possible to find some strange signals. By recording them, the paranormal researcher can make interesting discoveries.

Conducting experiments with signals picked up from the Earth is easy:

1. Place two metal bars (40 to 120 cm long) into the Earth with a separation of 3 to 40 m.
2. Connect them with any common wire to a transformer (to isolate the circuit from the signals and to match impedances). Any transformer with a low-impedance (low voltage) winding and a high-impedance (high voltage) winding can be used.
3. If the transformer has taps, add a switch that will allow you to choose the position on the windings that produces the best results.
4. Any audio amplifier with a power output ranging from 500 mW to 20 W can be used.
5. You can hear the sounds directly, or record them for later processing, to look for anything strange in the playback.

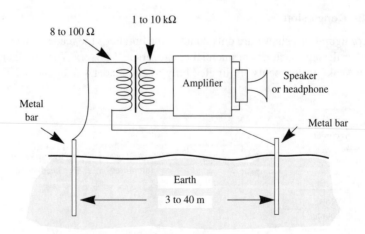

Figure 50 "Sounds from the Earth" pickup station.

6. Filters (noise, hum, etc.) can be installed between the transformer and the input of the amplifier for improved results. Figure 51 shows a "Sounds from the Earth" pickup station made with simple parts.

1.17.1 RF Experiments

So far, we have used only audible and ultrasonic sounds in our experiments, but they are not the only kinds of signals that can be used in experiments involving EVP. Many experimenters have found the voices using a radio receiver (MW, SW, or FM) tuned between stations or even using radio receivers with the tuning circuit adjusted in such a way that all the stations were mixed. The background noise generated by the circuit (by thermal agitation or picked up by the antenna as atmospheric noise) added to the mixed radio station signals can be used to reveal the voices.

It seems that all kinds of random signals, independent of their nature or frequency band, can support the stochastic resonance phenomenon and thereby be used to reveal the voices. The mechanism that causes this phenomenon is an interesting subject to be investigated.

It is not clear if the voices appear due to the noise produced by the loudspeaker of a radio receiver when tuned between stations, or if they arrive with the high-frequency noise tuned by the receiver. This is a question remaining to be investigated.

Experiments involving high-frequency signals start with two possible sources. One of them is radio transmitters or natural atmospheric or circuit noise. The other involves the use of transmitters that can fill the ambient with the necessary signals to support the EVP to be revealed.

Related experiments have used radio transmitters to produce electromagnetic signals to fill an ambient. These signals also can be used as carriers for the voices, which appear when demodulated by a receiver. Many researchers have related good results using 20 to 100 MHz transmitters. Experiments with very low frequency (VLF) transmitters can also be conducted.

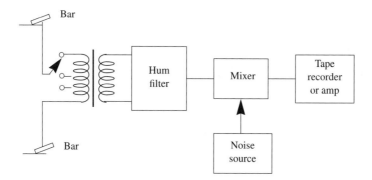

Figure 51 Another "Sounds from the Earth" pickup station.

Project 10: Noise Modulated Transmitter

An interesting circuit to be used by the experimenter is the one shown in Fig. 52. This circuit transmits a high-frequency (between 88 and 108 MHz) signal modulated by white (or pink) noise. The signal can be tuned by any FM receiver and transferred to a recorder, as shown in the same figure, or listened to directly by the researcher. The advantage of this circuit is that it is not necessary to find points between stations when tuning the receiver and trying to find the background noise necessary to the stochastic resonance effect.

The circuit produces a white/pink noise channel to be received by a common FM receiver. Placed near the receiver, its signal is strong enough to be tuned without any difficulty, even in regions with a congested FM band.

As explained before, many researchers believe that the noise in the electromagnetic band can be used more efficiently by the "beings" when producing their voices or other signals to be revealed to us.

The circuit shown has its own noise generator included, but any transmitter can be used with any of our noise generators in these experiments.

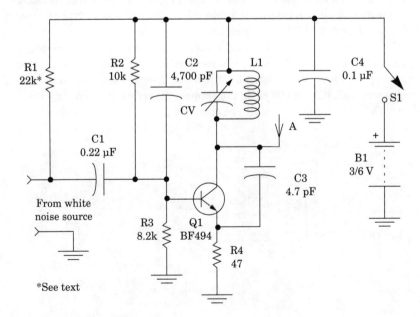

Figure 52 White noise transmitter.

How It Works

The modulation circuit is one of our white noise generators that use transistors. You can alter the project using other white noise generators as modulators.

The signal produced by this circuit is applied to a one-stage, low-power, high-frequency oscillator using a bipolar transistor. The output power of this stage is a few milliwatts, which can transmit the signals a distance of up to 20 m. This low power is important to avoid problems with the FCC, but it is high enough to allow us to pick up the signals using common FM receivers placed near the circuit. If larger antennas (rigid wire with length of 20 cm to 40 cm) are used, the signal can be picked up at distances of up to 50 m.

To tune the signal in the FM band, the coil is formed by four turns of 22 to 26 AWG wire in a coreless form with a diameter of 1 cm. The antenna is a piece of rigid wire 10 to 20 cm long. The signals can be tuned to an unused portion of the FM band.

Background noise also can be used to pick up images if a TV set is used as the receiver. A videocassette recorder can be used to register the images in a tape.

The snowy image in a TV is produced when weak signals are tuned in and the noise overlaps them, or when the TV is tuned between channels, in which case the snow is caused by noise picked up by the antenna (atmospheric noise) and also generated by the electrical circuits. Antique TV receivers are recommended for these experiments, as many new models have "squelch" circuits that cut off the signals when only noise is detected at the input.

Many researchers have recorded frames of snowy images that seem to contain no data but, when processed, can reveal shadows of persons and images of strange places or unknown beings. Several accessories can be used to separate the noise from the information, including electronic filters in the video recorder circuit or in the TV, optical filters, and signal processing circuits. (We will discuss the phenomenon as related to images in Part 2.)

It is important for the reader to understand that the researchers are trying to extract useful information from what we pick up as noise. We can draw a parallel with a situation in which a civilization has discovered radio but not how to separate the stations from one another. In such a case, the radio would pick up signals from many unknown stations, which would be mixed together and with background noise. The great problem we have and are trying to solve is how to precisely tune to the desired signals, separating them from the noise and other undesirable signals. And, most importantly, we need to determine their origin.

The circuit shown in Fig. 52 can also be used to send a signal to a TV tuned between channels 2 and 5, producing a snow pattern that can be recorded for research.

Assembly

The white noise transmitter can be mounted on a small printed circuit board as shown in Fig. 53. The white noise generator is mounted on a separate circuit board.

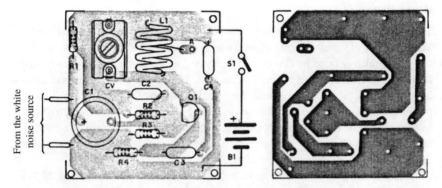

Figure 53 Printed circuit board for the white noise transmitter.

Any trimmer capacitor with a maximum capacitance ranging from 20 to 50 pF can be used. This component is used to tune the transmitter to a free point in the FM band or an unused TV channel between 2 and 6. The circuit is powered from 6 AA cells or a 9 V battery. Don't use any other type of capacitors where the parts list recommends ceramic ones.

Using the Circuit

Place the transmitter near the receiver. Tune the FM receiver to a free point in the band. Then, turn on S1 in the transmitter and adjust the trimmer until it picks up the strongest signal.

If your FM receiver system includes a tape recorder, you only have to record the signals and process them afterward. If not, you can place a tape recorder near the loudspeaker that produces the sounds.

Resistor R1 can be removed, in some cases. Conduct experiments to find the best value for this component (in the range of 22 kΩ to 1 MΩ).

In places where there are too many stations operating in the FM band, it can be difficult to find an unused frequency point to perform the experiments. In this case, you can reduce the size of the receiver's antenna so it cannot pick up signals from weak stations and place the transmitter closer to the radio.

Suggestions

- You can change the frequency of the transmitter by altering L1. Using 11 turns in the coil and replacing C3 with a 22 to 47 pF ceramic capacitor, you can tune to the signals of a CB receiver or transceiver.
- Use this circuit to transmit the signals to a receiver that will fill an ambient with white noise. Then use a microphone plugged into a tape recorder to pick up the ambient noise produced by the receiver. The voices then can be revealed.

Parts List: Project 10

Transmitter

Q1   BC494 or 2N2222, RF silicon NPN transistor

Resistors

R1   22 kΩ to 1 MΩ (see text)

R2   10 kΩ, 1/8 W, 5%—brown, black, orange

R3   8.2 kΩ, 1/8 W, 5%—gray, red, red

R4   47 Ω, 1/8 W, 5%—yellow, violet, black

Capacitors

C1   0/22 µF, ceramic or metal film

C2   4,700 pF, ceramic

C3   4.7 pF, ceramic

C4   0.1 µF, ceramic

CV   2 to 20 pF, trimmer capacitor (or any other)

Miscellaneous

L1   Coil (see text)

S1   SPST, toggle or slide switch

B1   3 or 6 V, two or four AA cells

A   Antenna, per text

Printed circuit board, cell holder, plastic box, wires, etc.

White Noise Generator

Use Project 2 printed circuit board and material

Project 11: Wireless Noise Generator (Medium Wave or Shortwave)

Many EVP experimenters use as the noise source a medium wave (MW) or shortwave (SW) receiver tuned between stations. The background noise, when reproduced by the receiver's loudspeaker, can fill an ambient and be used in experiments with tape recorders.

The background noise picked up by MW and SW receivers is due to the internal components of the circuit (thermal noise) and atmospheric phenomena (static discharge) with a level defined by many factors such as solar activity, atmospheric level of static electricity, ambient temperature, and others.

This background noise can be amplified with the circuit we show here. This circuit produces a high level of noise, extending the frequency range from the audible range into the shortwave band, as high as 20 MHz, and it can be picked up by any receiver placed nearby. Even in the VHF range, the noise can be picked up if the circuit is plugged directly to the antenna of a receiver. This allows the use of the circuit in experiments involving imaging with VHF TV.

Using the circuit is very simple: it is enough to tune any receiver (SW or MW) to a free point on the dial and place the noise generator near it. The noise will be transmitted and picked up by the receiver. The noise generator is powerful enough to send the signals to receivers placed up to 20 cm away.

How It Works

A white noise generator (one of the basic circuits described in this part) applies the signal to a power output stage using one transistor. The amplified signal is then applied to a ferrite core coil, which radiates it to receivers near the circuit.

As the load (ferrite coil) isn't a tuned circuit, the produced signal spreads throughout the band with an intensity that decreases with the frequency. The circuit is useful from the MW band (550 kHz) up to the SW band (20 MHz or 15 meter band) when transmitting the signals. But, if plugged directly to the antenna of a receiver, the signals can be tuned to frequencies up to 80 MHz.

The circuit can be powered using voltage sources from 9 to 12 V. In some cases, depending on the transistor used as noise source, the 9 V supply will not be enough to produce a good signal, and the voltage must be increased.

Assembly

Figure 54 shows the complete circuit of the noise generator. The coil is formed by 40 to 100 turns of any enameled wire between 22 and 30 AWG, or even plastic-covered wire.

The core is a ferrite rod with a length ranging from 10 to 25 cm. The core can be any convenient diameter.

A plastic box can be used to house the circuit. The coil can be installed as shown in Fig. 55. Dimensions of this plastic box are determined by the size of the cell holders. This figure also shows the circuit placement near a receiver when in operation.

If you intend to plug the circuit into a VHF receiver or TV set, a wire with an alligator clip can be connected to the collector of Q4. This clip is then attached to the antenna for transferring the noise.

Instrumental Transcommunication 63

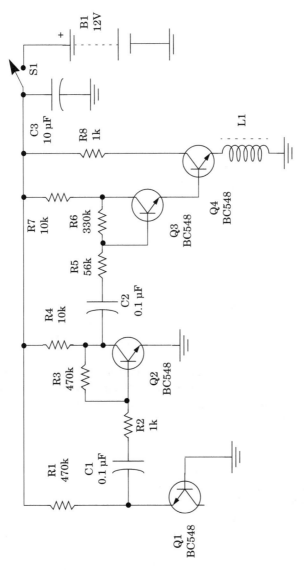

Figure 54 Wireless noise generator. This circuit sends noise to MW/SW receivers.

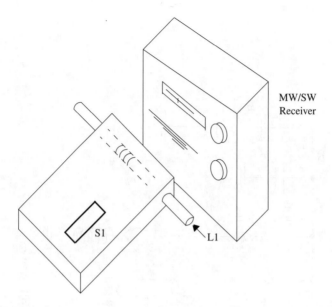

Figure 55 The circuit is housed in a plastic box as shown.

Using the Circuit

Place the generator near a MW or SW receiver. Tune the receiver to a free point on the band. The frequency choice is up to the experimenter. Many EVP experimenters, in their published works, provide a list of frequencies where they were successful in picking up signals.

Turn on the generator and find a position where the noise can be picked up easily. If the receiver is not sensitive enough, or if you are working in the high-frequency extreme of the SW band, you can plug the circuit directly into the receiver's antenna using the alligator clip.

Suggestions

- Alter the values of R1 and R3 to find the best performance.
- Replace Q4 with a BD135 and reduce R8 to 330 Ω or 470 Ω to increase the output power of the circuit.
- Wire a variable capacitor (120 to 470 pF) in parallel with the coil to tune the noise.
- Add a switch in series with a capacitor between 4,700 pF to 220 nF in the noise generator to produce pink noise, as suggested in Project 2. This circuit is placed between the base of Q1 and the ground.
- Use this circuit to increase the noise in a TV screen when performing experiments with images.

Parts List: Project 11

Semiconductors

Q1, Q2, Q3, Q4   BC 548, 2N2222 or any general-purpose NPN silicon transistor

Resistors

R1, R3   470 kΩ, 1/8 W, 5%—yellow, violet, yellow

R2, R8   1 kΩ, 1/8 W, 5%—brown, black, red

R4, R7   10 kΩ, 1/8 W, 5%—brown, black, orange

R5   56 kΩ, 1/8 W, 5%—green, blue, orange

R6   333 kΩ, 1/8 W, 5%—orange, orange, yellow

Capacitors

C1, C2   0.1 µF, ceramic or metal film

C3   10 µF/16 WVdc, electrolytic

Miscellaneous

L1   Ferrite coil (see text)

S1   SPST, toggle or slide switch

B1   9 or 12 V battery, power supply or 8 AA cells (see text)

Printed circuit board, plastic box, cell holders (two for 4 AA types) or battery holder, plastic box ferrite core, AWG wire (see text), wires, solder, etc.

Part 2
Experimenting with Images (Paranormal Images in Your TV)

> *Imagination is more important than knowledge. For knowledge is limited, whereas imagination embraces the entire world, stimulating progress, giving birth to evolution.*
>
> Albert Einstein

Not only voices can be recorded from paranormal sources as described in Part 1. The *electronic voice phenomenon (EVP)* can be extended to images, offering a new field of work for the paranormal researcher. This new kind of occurrence is called the *electronic image phenomenon (EIP)*.

The study of EIP began when some researchers taking photos of white walls with a particular degree of granulation discovered that the images revealed strange forms that were not originally visible. Under determinate conditions, the photos showed unknown shadows, strange forms, and even mysterious places immediately associated to paranormal sources as in the case of EVP.

Adding filters to the cameras, the photos increased in definition, and images with more details revealing strange beings and places were revealed. Again, the explanation of the phenomenon can start with the stochastic resonance, as in the case of EVP (see Part 1 for more details).

White light can be considered a type of noise, since it is formed by all frequencies of the visible spectrum, and the granulation of a white wall can produce some unknown kind of effect in the registered images. The phenomenon is also explained by many researchers as what they call "eidetic images." These images are associated with precise mental states found in children or immature adults and can have some correlation with the mental state of the researcher who is taking the photos or making experiments with TV cameras. As a result, it is believed, as a fundamental factor for success when registering the images, that the researcher must be in a special mental state.

A name that is frequently mentioned when talking about the origin of paranormal images registered in photos is Ted Serios. Ted Serios was an American, the son of Greeks, who could transfer to a photographic plate whatever he had in his mind, without any physical contact with the film or the camera. Researchers found that it was enough to place Ted Serios in front of a camera and, after a period of "concentration," a photo would be taken, and images from his mind would appear in it. Several important scientists observed these experiments.

As an example, consider an experiment in which the researchers tried to use a video tape to register the images. This experiment was made at the Denver Institute (Denver University) with the aid of Carl A. Hedberg (associate professor of the Electric Engineering Institute of Denver University), Ray M. Wainwriter (of the same institute), B. W. Wheeler (medical photography director of the Colorado School of Medicine), and Jules Eisenbud. Eisenbud discovered and, over a period of years, studied the special capabilities of Ted Serios. Eisenbud has published a book about Ted Serios that can be found at Amazon.com and other booksellers. (See the references at the end of this book.)

Ted Serios, with no explanation from the scientists who nevertheless discounted any possibility of tricks, also registered images of objects, places, and persons on magnetic tape. Taking into consideration that video images are placed on tape and decoded as a sequence of elementary fields rather than registered as a continuous grayscale pattern as in a photo, we really can't explain what kind of phenomenon was involved in this aspect of Ted Serios' abilities! But many paranormal events require the reader to come up with an explanation.

All the while, stories were surfacing of strange images picked up by television receivers in many places, all over the world, which resulted in increased interest in paranormal images tuned in this way. Dr. Max Berezovsky (Brazil) was one of the pioneers in this area of research. Starting with photographic experiments, he soon progressed to the use of the video casette recorder (VCR), a television receiver, and a camera. Focusing the camera on a TV screen tuned to a free channel, the images were registered on a casette tape. Between the camera and the TV screen, a filter was installed to "refine" the images. Using this process, and many miles of cassette tapes, Dr. Max Berezovsky registered hundreds of strange images involving places, persons, objects, and strange forms.

At the same time, many other researchers also registered images using the same basic configuration. Many of them have sites in the Internet where they present the results of their experiments. The results include images and, in many cases, also sounds.

The basic problem when working with these images is the same as related to sounds: the images are random, and we have no control over what will be registered. Since the conditions in which they appear are also not controllable, the researcher must count on a certain amount of luck to pick up some of them within just a few hours of experimentation.

We know that stochastic resonance is involved, as well as the difference between the synchronization of the TV image and the scan of the camera circuit, but what this means is something that remains to be investigated.

Mental condition, equipment used, and a certain number of favorable unknown conditions are additional factors that will determine what is picked up and when.

It is a fascinating field of research to be explored: how to "tune" the images so that we can choose their source, and how to get higher quality images.

2.1 The Basic Idea

The basic idea involved in these experiments is easy to understand. Again, stochastic resonance and white noise are involved but in a more "sophisticated" way.

When tuning a TV receiver to a free channel (in the VHF or UHF band, no matter where) the noise* appears in the video signal, producing an image with "sparkles" as shown in Figure 56. This sparkling image, formed by dark and clear dots and points, is determined by the intensity, duration, and polarity of the instantaneous value of the noise signal.

If this noise appears superimposed on a normal TV program with a very weak signal, the effect in the image is as if snow is falling over the scene, which explains why it is called *snow*. Figure 57 shows how the noise appears in the video signal and how it determines the pattern of the generated image. Depending on the level of the random noise signals, they produce clear or dark dots and points.

As dots and points are produced by a random signal, they change position in a fast, dynamic process. This means that the noisy image changes all the time.

Focusing a TV camera on the TV screen (we must consider the *scanning* process in this case, described in Section 2.3), the sparkling image is recorded. When edited, strange things can appear.

Researchers soon discovered that, due to the scanning processes and other factors, the images only appear in determinate frames. They also discovered that it is necessary to add filters to enhance the images.

Figure 58 shows an image registered by Dr. Max Berezovsky in his experiments using a TV camera, a VCR, a common TV receiver, and an optical filter. (We will explain in detail how to build filters later in this text.)

Many refinements were added to the original experiment by the researchers, such as the use of a computer to edit the images. Other possibilities include the use of many cameras simultaneously registering the same image, and editing them or mixing their signals for extended results.

Figure 56 Picture formed by small grains (dots and bars) due to atmospheric and circuit noise in a black-and-white TV.

---

* This noise is generated by atmospheric conditions or elements in the electronic circuitry, as mentioned in Part 1.

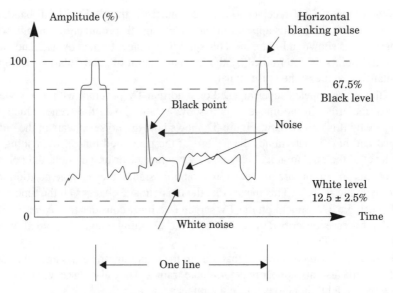

Figure 57 Signal corresponding to one line in a TV image.

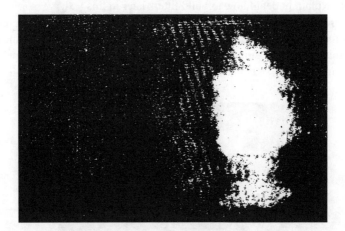

Figure 58 Strange image recorded by Dr. Max Berezovsky using one of the processes described in this book. (*Source:* courtesy of *Jornal da Tarde,* São Paulo.)

2.2 Experimenting with EIP

It is important to call the reader's attention to the fact that all of these fantastic experiments can be repeated using common equipment and parts. Plus, a reader who is experienced in electronics can add some refinements to the experiments, increasing the chance of new discoveries or to get better results when recording the images. The basic equipment necessary to begin the experiments, as listed below, is not expensive.

TV Receiver. A black-and-white or color TV is the first piece of equipment necessary to begin the experiments. The TV receiver will function as a sparkling image or noise source, much as white noise generators or the radio receivers function with the EVP. Any 14 to 29-inch or other TV receiver can be used. Some modern TV receivers have a "squelch circuit" that cuts the video signal when no station is tuned and fills the screen with a one-color picture (generally blue). This circuit is used to avoid the snowy image and white noise in the loudspeaker. This kind of TV receiver can't be used in these experiments. Old TVs that may no longer be useful for viewing common programs can be used in these experiments. Feel free to try that old TV set that has been abandoned in the garage to begin your experiments. It is not important whether the TV is a color unit. A black-and-white TV can be used successfully.

Camera. In principle, any TV camera can be used in the experiments with electronic image phenomena. Researchers recommend the use of the old model cameras that used tubes instead of the compact, modern cameras that use charge coupled devices (CCDs) as image sensors. The image tubes (vidicon or plumbicon) give the best results, as they use as their sensor a plate with a granulation pattern that seems to have some positive effect when recording the image. Maybe, in this case also, the structure of the sensor can influence the experiments—the effect has not been observed when using CCD cameras. A tripod is an important accessory for fixing the camera in front of the TV screen.

CCDs vs. Tubes. In the case of a tube-based video sensor, there is a sensitive plate inside the vacuum tube, and it is charged with a particular voltage. The electric charge of any part of the plate changes according the degree of light falling upon it. Focusing the image on the plate via a lens, the distribution of charges on the surface of the plate will correspond to the degree of illumination on each of its points.

An electron beam scans the plate, and the current picked from each point in the scan will be proportional to the degree of illumination generating the video signal. This video signal contains the information of the image, decomposed into lines and points.

In the case of a charge coupled device (CCD) sensor, there is a sensitive semiconductor plate formed by thousands or millions of small photo sensors, with each one representing one point of the image (a pixel). Projecting the image on it via a lens, each sensor will produce a current proportional to the degree of illumination at its image point. The video signal is then generated by activating the sensors in a sequence determined by a scan circuit.

VCR. The videocassette recorder (VCR) is a fundamental piece of equipment when experimenting with images. It is important that the VCR include a "cue" function. This is necessary when editing the images frame by frame. For experimenting with the VCR, the reader of course must obtain cables and all necessary equipment to connect it to the camera and TV receiver.

Filters. Optical filters can be used for experimental purposes by researchers. Use your imagination when creating devices suited to this task. Dr. Max Berezovsky used various metallic screens with small holes (uniform millimetric holes) as filters. A plastic or metallic sieve such as is found in chemical labs, or even in the kitchen, can be used as an optical filter. Placed in front of the camera, the sieve can reveal images when the tape is edited.

Polarizing filters, color filters (when working with color TV receivers), and even a piece of colored cellophane can be used as an experimental filter and combined to perform experiments.

Computer. A powerful tool that can be used by experimenters is the computer. Any image editor can be used to edit the images reproduced by the VCR. The great advantage to software filters is the wide range of capabilities they provide. The images, frame by frame, can be edited and transformed in ways that are not possible using the VCR only.

The algorithms found in many programs are powerful enough to reveal images where we ordinarily can't see anything. The computer can also be used to generate white noise or other kinds of noise on the monitor screen, replacing the TV receiver. Programs working with *fractals* are being used by many researchers to support experiments with images.

Miscellaneous. The experimenter should feel free to alter the basic, starting configurations suggested in the book for the experiments, and to add devices or equipment. Many changes can be made to upgrade your experiments, e.g.,

- Add some kind of light to the image. Fill the ambient with light from a source such as colored lamps or even infrared sources. Stroboscopic light sources can also be tested. Some researchers have obtained good results with photography by filling the ambient with infrared radiation. This could be applied to video.
- Try to pick up images not only from sources of noise such as the TV screen but from other origins such as a wall that is being illuminated by colored lamps, white lamps, stroboscopic sources, and other light sources.
- Mix the signal from the camera with signals generated by white noise sources.
- Add sounds to the experiments via the audio channel of the VCR to make simultaneous EVP tests.
- Use the monitor of a PC as noise source when making the experiments. Software that works with noise, and even fractals, provides a very interesting approach when setting up the experiments.

- Use the photographic camera or the printer (when working with the computer) to have images in paper. The images on paper can also be transferred to the computer using a scanner if you don't have the appropriate hardware to transfer images directly from the VCR.
- Add to your equipment the kit to reveal photography. A dark room is necessary when you intend to reveal your photos at home. There are kits to reveal photos including all the necessary material and books teaching how to do it.

2.2.1 Procedures

When working with images, the recommended procedures described in the experiments with voices (EVP) are also valid. The reader must pay special attention to the concept of *interpretation* described in Section 1.10. And, as in the case of EVP, it is important to avoid *projection* (described in the same section).

With images, there is a natural human tendency to see something that is consistent with our expectations, which can create the illusion that we did indeed see something more than what is actually present in an image. By this process, we might see only a single shadow and project it into a complete image that is not in fact present. Our brain attempts to make something logical out of the illogical. This is very dangerous to the serious experimenter who wants to find the real facts involved in the EIP and not be tricked into false conclusions. We remember again that, although this book is about paranormal experiments, it focuses on the scientific side of them. The idea is to help the hobbyist, amateur experimenter, and serious experimenter to use electronic equipment to determine what is real and what is not real in all subjects of paranormal science addressed in this book.

Again, we recommend that the reader take much care when processing the images and all the information collected in the experiments. Not every shadow or line is going to make sense. Not every image will be understood. Keep this in mind to avoid tricking yourself.

If the reader wants additional information, we can recommend a brief look at scientific papers about "eidetic images" and "stochastic resonance." These provide additional information about the influence of our mind on the interpretation of the phenomena.

2.3 Scanning, Frames, and the TV Image

The use of a VCR, computer, and TV receiver involves the production of images in a process that the researcher must understand to make correct interpretations of the phenomena. Readers experienced in electronics probably know how a TV works, and so the terms *scanning, frame, field,* and others are very familiar.

To avoid false interpretations about the origin of strange images picked up in an experiment, and to better understand how they are produced, some information about how a TV image is generated is important. A brief explanation is given next.

The image on the screen of a TV receiver is produced by a process in which an electron beam draws a complete image, point by point, by striking a phosphor coating on the face of the picture tube (i.e., the screen of the cathode ray tube or CRT).

To recreate a complex scene that occurs simultaneously at many points in space but is transmitted as a sequential stream of information, a scanning process is used. Scanning begins at the top left-hand corner of the screen, and the first line is swept rightward across it. The formed trace is completed when the right side is reached. At this moment, the *retrace* begins, during which time the electron gun returns to the left-hand side of the screen, and no picture information is transmitted. The second line is then produced below the first line in the same manner, and the process continues. After the bottom line is traced, the beam is again turned off and is repositioned at the top of the screen.

We can compare this process to that followed in reading a book, where the reader's eyes start at the upper left hand corner and the information is "picked up" only during the left-to-right scans.

In North American television, 262.5 lines are transmitted and reproduced in each scanning cycle. These 265.5 correspond to a *field,* but they don't constitute the complete scene, which needs 525 lines and two scans. The complete scene corresponds to a *frame*. By this process, 30 frames per second are reproduced on the screen, corresponding to 60 fields.

This means that the total lines produced each second, known as the *horizontal scanning rate,* is 15,750 per second. This means that the horizontal frequency of a TV is 15,750 kHz.

Note that our eyes do not process images quickly enough to see the electron beam scanning the screen, and the effect is that of a complete image. Even when the frames change, our eyes are not fast enough to see the change, giving the impression of movement in the image.

To recreate a television picture, pulses to synchronize the precise timing of the scanning are introduced into the lags between parts of the video signal at the TV camera. Two pulses are inserted in the signal: the horizontal pulse, which is needed to determine when each line ends and begins, and the vertical pulse, which tells the TV circuit when a field begins and ends. These pulses are called *synchronizing* pulses, and they determine the stability of the reproduced image. If the TV receiver loses the pulses, the image rolls and disintegrates. The typical TV signal is then formed as shown in Fig. 59.

Between the synchronization pulses, we find the video information corresponding to one line of the image. Figure 59 also shows how a noise is placed in the video signal, corresponding to one line of an image. It is very important for the reader to keep in mind this process of image formation when trying to explain any mysterious image picked up in the experiments.

Color TV. In a color TV, the screen is covered by phosphor points of the three basic colors (red, green, and blue—RGB). Three electron beams scan the screen,

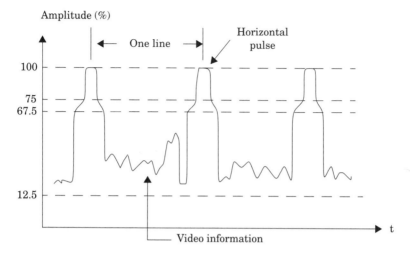

Figure 59 The television signal.

with each of them determining how much of each color is applied to form a point of the image of a determinate color.

Beat. Another important phenomenon to keep in mind when picking up images with a video camera from a TV receiver is that the scan frequencies of the video camera and the TV receiver are not exactly the same, and the differences can result in some unwanted effects when editing the images. It may happen that, at the instant in which the image picked up by the camera is in the first scan line, the television receiver is reproducing the last line of the same frame. This means that, even considering that the line persists on the screen for a certain time interval, the picked up line is not as clear as it could be. The basic parameters of television are shown in the table below.

Basic Parameters of Television

Parameter	Description	Value
Scan line number	This number determines the vertical resolution. A finer image results with an increase in the number of scan lines.	525 lines (63.5 μs per line)
Field frequency	This number determines the amount of flicker. The greater is the field frequency the greater is the rate at which pictures are produced in the screen.	60 per second (60 Hz) 525 lines 33.4 ms per frame
Frame frequency	As we have two fields per frame, the frame frequency is half the field frequency.	30 per second (30 Hz) 525 lines 33.4 ms per frame

Observation: If the reader isn't in the U.S.A., it is important to know exactly the characteristics of the TV system used in your country, which may differ from those listed above. This can be necessary to interpret any paranormal phenomena involving images without the risk of mistakes.

Photography. These images may be captured using a standard camera as well, after which they can be scanned and further processed. Many researchers have obtained good results in experiments with photographic cameras aimed at white walls illuminated by different light sources, and using filters. The technique is the same as described for TV.

In this case, a Polaroid or standard camera is recommended. Digital cameras are less desirable, as they use CCD sensors that narrow the noise bandwidth and reduce the probability of the stochastic resonance effect appearing. In some cases, digital cameras also present difficulties in the use of optical filters. Here too, electronic circuits can be used to produce white noise in the experiments.

2.4 Practical Circuits

Several practical circuits can be built by the EIP experimenter to improve the experimental procedures. In the following text, we present some examples of projects that use common parts and can be added to the basic configurations for improved results. These include noise generators, light effects, noise transmitters, and others.

Since the circuit specifications are not critical, many variations can be made to change the performance and thereby get better results. We also should note that these circuits can be used in other experiments involving paranormal phenomena.

Project 12: Wireless Sparkling Image Generator

The circuit shown in Project 11 (page 61) can be used to generate a sparkling image (drizzle image) in a common VHF TV receiver. The white noise signal produced by the circuit can be used at frequencies up to 70 MHz, so it is recommended that you tune the TV receiver within the low VHF channels (between channels 2 and 6).

As shown in Fig. 60, the circuit is plugged to the antenna terminals or the telescopic antenna using an alligator clip or terminals. The television receiver is tuned to a free channel between 2 and 4 for best results, but the signal can also be picked up in channels 6 and 7.

The sparkling or drizzle image produced by this circuit is better than natural atmospheric or circuit noise, as the noise produced by the circuit is more powerful. Frequencies used in VHF TV channels are shown in the table that follows.

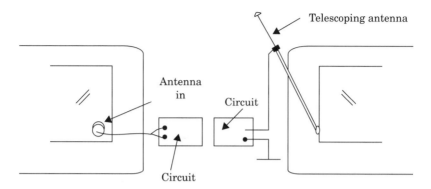

Figure 60 Plugging the circuit into a TV receiver.

Frequencies Used in VHF TV Channels

Channel	Frequency Band (MHz)	Video Carrier (MHz)	Sound Carrier (MHz)
2	54–60	55.25	59.75
3	60–66	61.25	65.75
4	66–72	67.25	71.75
5	76–82	77.25	81.75
6	82–88	83.25	87.75
7	174–180	175.25	179.75
8	180–186	181.25	185.75
9	186–192	187.25	191.75
10	192–198	193.25	197.75
11	198–204	193.25	203.75
12	204–210	199.25	209.75
13	210–216	205.25	215.75

Project 13: TV White Noise Generator

The next circuit can transmit noise to any TV receiver placed near it. The range is approximately 20 m, and the signals can be picked within any free channel between Channels 2 and 6. Alterations can be made to the circuit to allow operation in the high channels between 7 and 13.

The noise produced by this circuit can be used in experiments to pick up the images using a camera and the TV as shown in Fig. 61. When picking up images from a TV using a camera, the reader must take into consideration the beat phenomenon that occurs due to the different sweep frequencies of the two devices. This means that when the TV is producing one line on the screen, the camera is registering another line. This difference in synchronization causes the images recorded from a computer monitor or other TV set to be chopped when reproduced. Consider this fact when editing the image frame by frame.

How It Works

The transmitter is a simple low-power VHF oscillator using an NPN RF silicon transistor. The transistor is wired in the common base configuration. L1/CV form the resonant circuit determining the frequency of operation of the transmitter.

With the coil indicated in the project, the circuit will oscillate between 54 and 88 MHz, corresponding to channels between 2 and 6, but the coil can be altered to make the circuit oscillate between 174 and 216 MHz, corresponding to channels 7 to 13. See in the suggestions how to make this alteration.

Bias is given by R1 and R2 and feedback by C3. The noise is applied to the base of the transistor by C1. Any of the noise generators described in this book can be used as signal source to modulate this transmitter.

It is important to see that the level of noise determines how the signal produced by the transmitter spreads by the channel band. If enough noise is applied to the input, the transmitter will also produce noise in the audio band, allowing the experimenter to perform recordings in the audio band of the tape. Parallel experiments with EVP can be performed this way.

The power supply is formed by four AA cells or a 9 V battery. Since power consumption is not high, a 9 V battery can power the circuit for several hours.

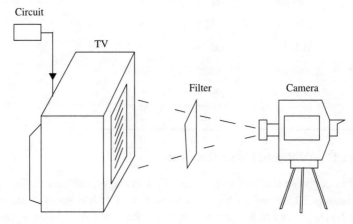

Figure 61 Using the circuit.

Experimenting with Images 79

The antenna is a small piece of rigid wire 10 to 30 cm long. Do not use long antennas, as they may interfere with other TV sets in your neighborhood. *Doing so can create severe problems with the FCC!*

Assembly

A diagram of the TV transmitter is shown in Fig. 62. The transmitter can be mounted on a small printed circuit board as shown in Fig. 63. Since the circuit is not critical, mounting can also be accomplished using a small terminal strip as shown in Fig. 64.

Equivalents of the specified transistor can be used, such as the 2N2222 or 2N2218. But observe that the terminal placement is different from the original.

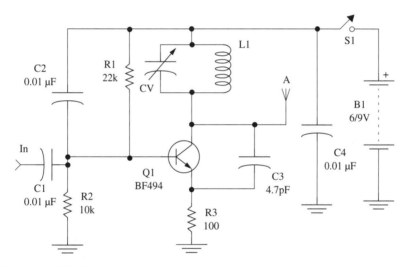

Figure 62 TV white noise generator.

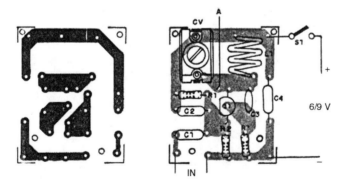

Figure 63 Printed circuit board used for the TV transmitter.

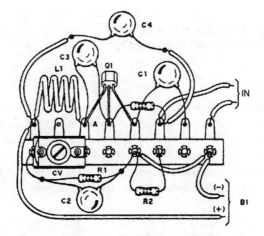

Figure 64 A terminal strip can be used as the chassis.

Both mountings can be placed inside a plastic box with a jack for the input signal and a hole for the antenna. The coil is formed by four turns of 22 or 24 AWG bare wire with a diameter of 1 cm with no core (air core). This coil will produce signals between channels 2 and 6.

Any trimmer capacitor with maximum capacitances between 15 and 50 pF can be used. The capacitors indicated as ceramic types must not be replaced by other types.

Using the Transmitter

The first option is to place the transmitter near a TV receiver (see text) tuned to a free channel between 2 and 6, as shown in Fig. 65 and use a camera to pick up the images from the TV screen. The second option is shown in Fig. 66. In this case, a video casette recorder is used as tuner and placed near the transmitter. The video casette is adjusted to a free channel between 2 and 6. In both cases, adjust CV to select the signal from the transmitter making the recordings.

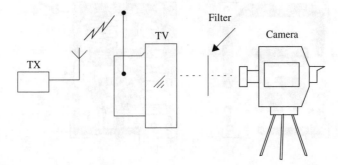

Figure 65 The TV is tuned to a free channel between 2 and 6.

Experimenting with Images   81

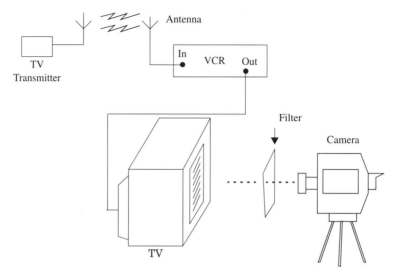

Figure 66   Second option for use of the TV transmitter.

WARNING: Do not tune the circuit to the frequency of channels that are in use in your location. The transmitter will interfere with any TV placed near the transmitter when tuned to the same channel.

Be sure that, in the channel you are using to pick up the images, there is no signal from distant stations. A weak signal superimposed on the noise can be confused as a "paranormal" image. You would not like to have the face of your favorite TV show's announcer included in a next-dimension photo! Unplug the external antenna to avoid these signals.

Suggestions

- Use a two-turn coil and replace C3 with a 1 pF capacitor. The transmitter will produce signals between channels 7 and 13.
- A "gray" image can be produced if the noise source of the circuit is disabled. Some experiments can be performed using this image. In this case, the circuit will generate only a pure video carrier, filling the screen with a uniform image. The brightness and contrast can be adjusted using the TV controls.
- If a sound source (audio oscillator or the output of a tape recorder with an EVP tape) is plugged into the input of this transmitter, it will modulate the image with unexpected effects that can be used in experiments.
- A low-frequency oscillator plugged to the input of the transmitter will produce horizontal bars on the screen. They can be used in experiments.

Parts List: Project 13

Semiconductors

Q1   BF494 or equivalent RF NPN silicon transistor

Resistors

R1   22 kΩ, 1/8 W, 5%—red, red orange

R2   10 kΩ, 1/8 W, 5%—brown, black, orange

R3   100 Ω, 1/8 W, 5%—brown, black, brown

Capacitors

C1   0.01 µF, ceramic or metal film

C2   0.01 µF, ceramic

C3   4.7 pF, ceramic

CV   Trimmer capacitor (see text)

Miscellaneous

L1   Coil (see text)

S1   SPST, toggle or slide switch

B1   6 or 9 V, four AA cells or one 9 V battery

A   Antenna (see text)

PCB or terminal strip, plastic box, battery clip or holder, wires, solder, etc.

Project 14: Horizontal Bar Generator

Readers can make experiments with kinds of images other than a sparkling image. The circuit described here produces a horizontal bar pattern, filling the TV screen, and can be used for dynamic effects when recorded from a camera due to the beat effect of the scanning frequencies of the two devices (TV and camera). Figure 67 shows the image generated by this circuit on a TV screen. This circuit includes the low-frequency oscillator suggested in the previous project to produce the horizontal bars in a TV screen.

The experimenter can adjust the horizontal synchronization of the TV receiver to obtain a pattern like the one shown in Fig. 68. With the aid of optical filters, the received images can be processed to reveal some interesting things.

How It Works

Our circuit is formed by a single low-frequency generator that applies the signal to a video carrier. The circuit is tuned to a free channel in the VHF TV band.

Experimenting with Images 83

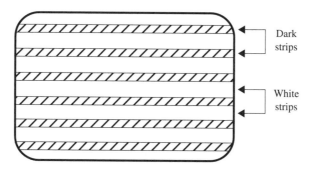

Figure 67 Image generated by the described circuit.

Figure 68 Adjusting the horizontal synchronization control, we can produce an image like this one. Recording the image with the aid of a camera and editing it, strange images can be found.

When picked up by the TV, the video circuits see the low-frequency component of the signal as image signals, converting them into horizontal dark and clear lines according to the logic level. The number of lines and their width depend on the low-frequency oscillator and the duty cycle of the signal. These values can be adjusted by two controls: P1 and P2 simultaneously control the duty cycle and therefore the width of the dark and clear lines that appear on the TV screen. Of course, the lines are not in color but, for this experiment, they are suitable.

The low-frequency circuit is made with one of the four NAND gates of a 4093 CMOS IC. The other three gates are used as a digital amplifier, buffering the signal to a high-frequency stage using a transistor. The high-frequency stage of the circuit is a low-power transmitter that sends the signals to a TV placed up to 10 m away. Therefore, no physical connection is needed between them.

CV determines the operation frequency and must be adjusted to a free channel in the low VHF TV band, between channels 2 and 6. L can be altered as in the previous project to transmit into the upper VHF channels between 7 and 13.

Assembly

The complete schematic diagram of the Horizontal Bar Generator is shown in Fig. 69. The components can be placed on a small printed circuit board as shown by Fig. 70.

Any trimmer capacitor with capacitance in the range between 1–15 and 3–30 pF can be used in the project. The coil is formed by four turns of 20 to 24 AWG wire with a diameter of 1 cm without core (air core).

The transistor in the original project is the BF494, but you can use equivalents such as the 2N2222, 2N2218, etc. You only have to take care with the terminal placement. The 2N2222, for instance, has a terminal placement different from the original BF494.

Without an antenna, the circuit can send the signals distances up to 10 m, but you can use a small antenna to extend this range to up to 30 m. The antenna is a piece of rigid wire 10 to 30 cm long connected to the collector of transistor Q1.

To power the circuit, two or four AA cells can be used. The current drain is not high, and the cells' life can extend to several weeks in normal use. All the components can be fit into a plastic box, making the unit portable and easy to use.

Testing and Using the Circuit

Place the circuit near a TV tuned to a free channel between 2 and 6. Turn on the power supply and adjust CV until the signal from the bar generator is picked up. Adjust P1 and P2 to get the desired bar pattern on the screen.

The sound must be at minimum, as the signal from the generator is reproduced by the TV loudspeaker as a continuous tone. Experiments can be made with the bar pattern on the screen, but you can add some "effects." Adjusting the horizontal synchronization control, you can bend the picture and get a random pattern as shown before.

Suggestions

- If the circuit is placed too far from the TV, the signal will not be strong enough to cover the natural noise. Therefore, you will have an image composed of only the horizontal lines and the snowy image of a free channel. Be sure to not use a channel that includes signals from distant stations that can mix with your signal, allowing the weak images to affect the experiment.
- Vary C1 in the range between 1 nF and 220 nF to change the line pattern.
- Place a mixer between C2 and the transmitter stage. Apply the signal of any white noise source to the other input of the mixer. You can work simultaneously with the bar pattern and the noise pattern generated by the two circuits.

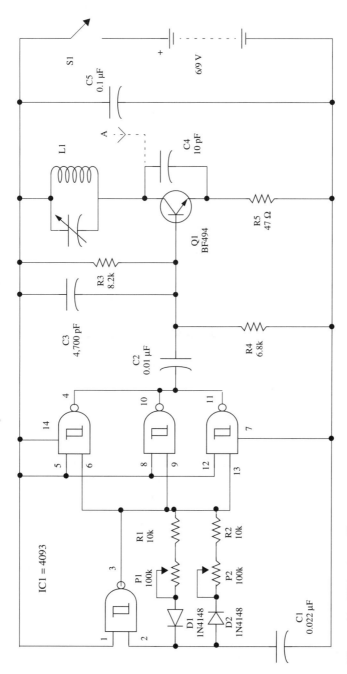

Figure 69 TV horizontal bar generator.

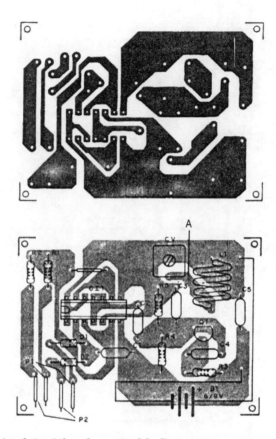

Figure 70 Printed circuit board suggested for Project 14.

- Replace the transistor with a UHF type such as the BF689K or BF967 (Phillips), reduce C4 to 1.0 to 2.2 pF, and reduce the coil to one turn with the same diameter, and the circuit will produce signals in the UHF low band (channels from 14 to 40).
- A modulation control can be added between C2 and the input of a transmitter. Replace R4 with a 10 kΩ potentiometer and wire C2 to its cursor.

Parts List: Project 14

Semiconductors

IC1	4093 CMOS integrated circuit
Q1	BF494, 2N2218, or 2N2222 NPN high-frequency silicon transistor
D1, D2	1N4148 or 1N914, general-purpose silicon diodes

Resistors

R1, R2	10 kΩ, 1/8 W, 5%—brown, black, orange
R3	8,200 Ω, 1/8 W, 5%—gray, red, red
R4	6,800 Ω, 1/8 W, 5%—blue, gray, red
R5	47 Ω, 1/8 W, 5%—yellow, violet, black

Capacitors

C1	0.022 µF, ceramic or metal film
C2	0.01 µF, ceramic
C3	4,700 pF, ceramic
C4	10 pF, ceramic
C5	0.1 µF, ceramic
CV	Trimmer capacitor (see text)

Miscellaneous

P1, P2	100 kΩ potentiometers
L1	Coil (see text)
S1	SPST. toggle or slide switch
B1	3 to 6 V, two or four AA cells

PCB, plastic box, antenna (see text), AA cell holder, wires, solder, etc.

Project 15: Video Inverter

Some paranormal images registered in the experiments can be revealed only if inverted (taken from the negative to positive form), since they are negative pictures. They appear like the negative of a photograph due to the inversion of the polarity of the video signal recorded on tape. This condition can complicate the researcher's works when trying to identify the image.

Using a video inverter, the researcher can "uninvert" images in the negative form, turning them into a positive form as shown in Fig. 71. To work with these images, the EIP researcher will find a video inverter to be a useful tool.

A video inverter that can be placed between the VCR and the TV is described in this project. This circuit can also be placed between a camera and the VCR, registering inverted or negative images, adding a new kind of experiment to the researcher's arsenal. Other paranormal experiments can be performed using this circuit, e.g., involving extra sensory perception (ESP) or transcendental meditation.

Figure 71 Images can be transposed to negative form as shown by the above photos of Ted Serios (positive on left, negative on right).

How It Works

The circuit uses only common transistor stages. The amplitude of the input signals is controlled by P1. By adjusting this control, we can find the correct value of the input voltage to avoid saturating the circuit and to give to the pulse trigger circuit the signal needed for correct operation.

The video signal is inverted by Q1. This transistor is wired in the common emitter configuration where the input signal applied to the base appears with its phase inverted at the collector.

From this stage, the signal is applied to Q4, which is connected in the common collector configuration. In this configuration, the phase of the amplified signal is not inverted and, after amplification, we obtain it from a low-impedance output signal.

The synchronization pulse is operated by Q2 and Q3. These transistors amplify the pulse and mix it with the video signal, applying it to the base of Q4.

P2 is adjusted to find the best point of signal inversion, determining the contrast of the picture in the TV screen. S2 is used to return the signal to the original phase. This switch can be used to compare the inverted signal with the original signal when editing an image. The LED is a "power on" indicator and can be omitted.

The circuit can be powered from AA cells or a power supply plugged to the ac power line. The current drain is very low, extending the cells' life to many weeks. Originally, this circuit was designed to descramble TV signals but, as the reader can see, there are many other uses for this simple configuration.

Assembly

Figure 72 shows the diagram of the video inverter. The components can be mounted on a printed circuit board as shown in Fig. 73.

The circuit is not critical, but it is important to keep all the connections short. Cables with RCA plugs can be used to connect the video inverter between the VCR and the TV receiver (video input). Choose good quality, short video cables to avoid signal losses and distortions. Distortions can mask the results of the ex-

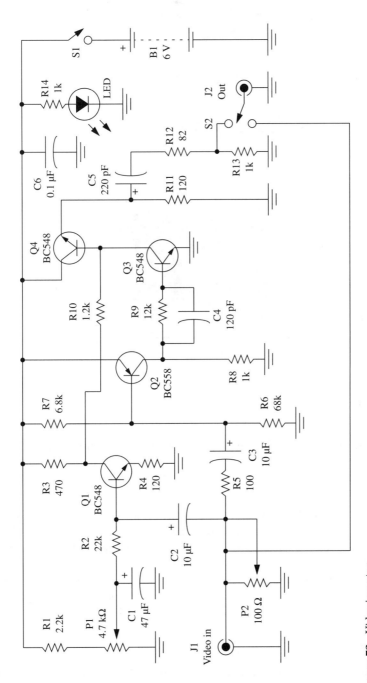

Figure 72 Video inverter.

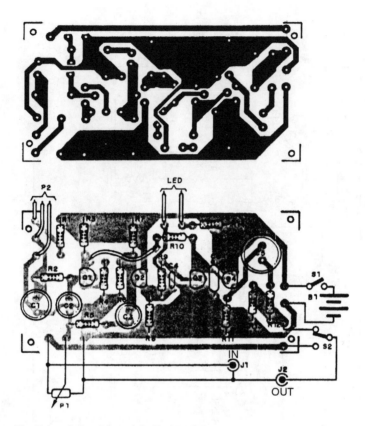

Figure 73 Printed circuit board for Project 15.

periments. All the parts including the power supply can be housed in a small plastic box.

Using the Circuit

The circuit is placed between the VCR or camera and the TV receiver as shown in Fig. 74. P1 and P2 must be adjusted to give a stable picture on the screen. If you don't obtain a stable picture in the TV, you may need to experiment with different values of C1.

Figure 75 shows how to perform a experiment with images taped from a screen and illuminated by several kinds of light sources while monitoring them with a TV receiver. The device used between the camera and the VCR/TV is a *video separator.*

Suggestions

- All experiments involving images taped from white walls or a TV tuned to a free channel also can be made using the video inverter to modify the image.

Experimenting with Images 91

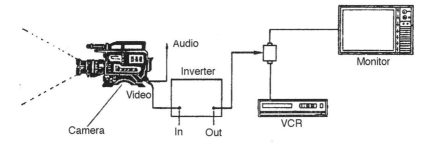

Figure 74 Using the inverter.

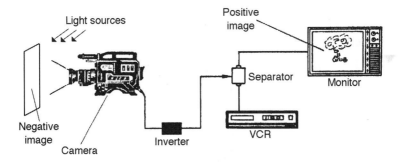

Figure 75 Experimenting with negative images.

- Capture the images in the normal and inverted (negative) modes and edit them using a computer. Print the two images and compare.
- Use the video inverter to produce an inverted white noise image in a TV, performing the common experiments with EIP.
- Perform experiments using this circuit and the bar generator of the previous project.

Parts List: Project 15

Semiconductors

Q1 to Q4   BC548, 2N2222 or equivalent, general-purpose NPN silicon transistors

LED1   Common red LED (optional)

Resistors

R1   2.2 kΩ, 1/8 W, 5%—red, red, red

R2   22 kΩ, 1/8 W, 5%—red, red, orange

R3   470 Ω, 1/8 W, 5%—yellow, violet, brown

R4	120 Ω, 1/8 W, 5%—brown, red, brown
R5	100 Ω, 1/8 W, 5%—brown, black, brown
R6	68 kΩ, 1/8 W, 5%—blue, gray, orange
R7	6.8 kΩ, 1.8 W, 5%—blue, gray, red
R8, R13, R14	1 kΩ, 1/8 W, 5%—brown, black, red
R9	12 kΩ, 1.8 W, 5%—brown, red, orange
R10	1.2 kΩ 1/8 W, 5%—brown, red, red
R11	120 Ω, 1/8 W, 5%– brown, red, brown
R12	82 Ω, 1/8 W, 5%—gray, red, black

Capacitors

C1	47 μF/12 WVDC, electrolytic
C2, C3	10 μF/12 WVDC, electrolytic
C4	120 pF, ceramic
C5	220 μF/12 WVDC, electrolytic
C6	100 μF/12 WVDC, electrolytic

Miscellaneous

P1	100 Ω potentiometer (linear or log)
P2	1 kΩ potentiometer (linear or log)
S1	SPST, toggle or slide switch
S2	SPDT, toggle or slide switch
B1	6 V, four AA cells or power supply (see text)
J1, J2	Input and output RCA jacks

Printed circuit board, plastic box, shielded cable, RCA jacks and plugs, cell holder, knobs for P1 and P2, wires, solder, etc.

2.5 Suggested Experiments

Several configurations that combine VCRs, cameras, and a TV with the previously described projects can be suggested. In Fig. 76, we show some of them.

- In Fig. 76a, we have the basic configuration to pick up paranormal images with a TV camera. The TV receiver tunes white noise, producing a "sparkling" image that is picked up by a camera aided by several types of filters as described before. The experimenter is free to try various types of filters.

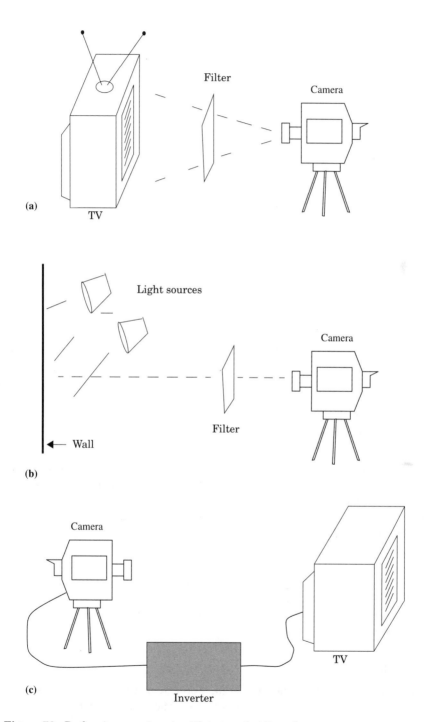

Figure 76 Performing experiments with images (continues).

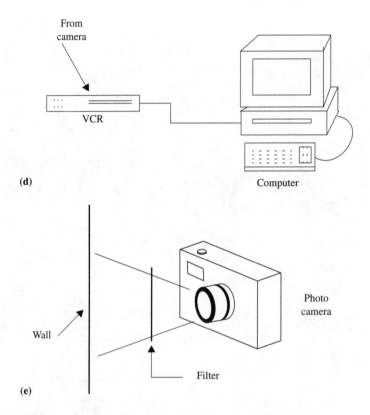

Figure 76 (continued) Performing experiments with images.

- In Fig. 76b, we show the camera pointing at a white wall that is illuminated by several types of light sources: white light, colored light, infrared, modulated light, etc. This camera can also be equipped with filters.
- In Fig. 76c, we see the signal picked up by the camera being processed by a video inverter before it is sent to a TV receiver.
- In Fig. 76d, the received signal is applied to a computer. The images can be edited more easily with this resource.
- Fig. 76e shows the use of a photographic camera to pick up the images. The photo can be scanned and edited with a computer. Filters can be added to the experiment as in the other cases. Various light sources can be created by the experimenter.

2.5.1 Interpreting the Results

Hundreds of researchers from all over the world are now trying to pick up sounds and images from the next dimension—or from wherever they come. The correct configuration to find what they seek is not known. Many researchers claim to have found the solution and maintain that their equipment is the best. Some sell equipment for hundreds or thousands of dollars (although it may not be even as sophisticated as the devices we describe here) and attempt to convince the purchaser that this equipment includes some special scientific secret.

Many photos and images are displayed in the media. For a sample, simply type "kirlian" into any search engine on the Internet. But when paranormal equipment vendors are asked for details about the devices they are using or are invited to demonstrate and explain their operation, they become evasive, saying that their equipment involves "secret techniques" or that the principles are too difficult to explain to "common people" (possibly in an attempt to justify the high prices).

In practice, the interpretation may also vary according to the beliefs of each researcher, according to what he wants people to see in the images. Extrapolation and radicalization of results, as in the case of EVP and EIP, is also a common occurrence. We warn the serious researcher to be very careful when interpreting the results and describing his discoveries in this fantastic field.

2.5.2 The Computer

Researchers have found that any electronic circuit is sensitive to next-dimension signals and paranormal phenomena and, under certain conditions, can be used to pick them up. Under special conditions, unexplained signals can cause an unexpected brightness variation in a domestic lamp; interference in audio or radio equipment, cellular telephones, fax machines, and auto electrical systems; and many other strange occurrences. Even computers aren't immune from this influence.

Some experimenters have documented signals being picked up by a common computer and transferred to a printer without any involvement by the human operator. Strange messages, signed by people who died many years before or by unknown beings, have been registered this way and reported by researchers!

How to induce a computer to do that is an interesting problem that remains to be discussed and solved by the paranormal researchers. Some kind of device should be invented that can instigate this process. This is perhaps an interesting field to be considered by readers of this book.

Another interesting concept to be considered is how one might connect a data acquisition circuit board (analog-to-digital converter, or ADC) into the serial or parallel port of a computer, and what kind of software program could be developed to process the data. The signal applied to the ADC could be from a tape recorder (EVP) or even from white noise generators such as the ones described here, allowing the signals to be processed in real time.

The computer can also be used for other purposes when working with the paranormal voices or images. Although the microphone input of a computer can be used directly to record signals, with some limitations it can also be used as an input port for radio receivers, white noise generators, and other noise-producing devices.

To transfer the signal from a tape to a computer without receiving any noise from the room, it is necessary to use a shielded audio cable (available at any electronic supply store). The reader must be sure to purchase one with connectors that will fit properly into the devices to be used. If in doubt, tell the salesman that you intend to record on your computer from a small recorder. The salesman will probably know what size connector you will need based on that information.

Once you have found the correct cable, you simply run the cable from the output of the tape deck to the microphone input on the PC. Many audio processing software packages can be used to edit the taped signals. However, if you don't want to invest money in sophisticated software for this task, the resources of Windows 95 or 98 or the Mac OS are sufficient for basic recording.

Using Windows

The procedure to transfer the sound using Windows 95 is as follows:

1. Once the wiring is in place, go to the "start" button and from there to the accessories.
2. In the accessories, find the multimedia icon and click on it.
3. The next step is click on the sound recorder icon. Opening this device, you must set the recording option for maximum quality. Go to the edit pull-down menu at the top of the recorder and select audio properties. You'll see a window with an upper and lower section in it.
4. The playback device is shown at the top, and the recording device is shown at the bottom. In the recording area appears a little window titled "preferred quality." Set this window to "CD quality." If this is not an option in your computer, select one with at least 16 bit, 22 kHz mono characteristics. This will enable the system to retain the same quality as the tape used as signal source.
5. Going now to the file pull-down menu at the top of the tape recorder, select "New." With this selection, the red record button will be activated and ready for use.
6. Set the tape to begin a few seconds from the beginning point of the signals you want to transfer. Point the mouse to the record button and press the "play" button on the tape recorder, letting the tape run. You will hear the recorded sound as it comes into the computer. Wait a few seconds and press the "record" button on the computer to save the sound to the hard disk. You may need to experiment a bit to synchronize the operations (playback of the tape and recording into the computer) properly.
7. When the recording is done, press the "stop" button and rewind the virtual tape back to the beginning.

Playing Back the Sounds from the Computer

- Press the "play" button (arrow pointing toward the right) and review your recording. Adjust the volume control to get a high enough volume to find the

voices. If the voices are very faint (in the beginning of the experiments, the results may be not so good), you have to keep trying.
- If your playback program has a graphic equalizer, you can use it to cut the bass and treble (low and high frequencies), giving emphasis to the middle, where voice frequencies are concentrated.
- If you are unsure about how to operate the program, go to the "help" section to see what is going wrong.

The main limitation when using the PC is that, in the process of converting the signal from analog to digital format, part of the high-frequency spectrum, where white noise is predominant and where important information can be overlapped, is lost. This can affect the results. The advantage is that the computer software can be used to examine the signal in detail and even to detect information that cannot otherwise be accessed.

A good approach is to try the software used for the analysis of space signals in the Seti Project (http://www.setileague.org), as described in the July/1999 issue of *Popular Electronics* (http://www.gernsback.com). This software can be adapted to separate signals from noise in your research.

2.5.3 More Information about ITC and EVP

There are many resources of ITC and the EVP in the Internet. As the Internet is dynamic, and every day pages and sites are changed or disappear without any notice, the best way to find information in the World Wide Web is by using the search engines such as Yahoo, AltaVista, Lycos, Infoseek, Hotbot, and others. You simply enter the keyword and investigate the hundreds or thousands of documents that are returned from your inquiry.

Some suggested keywords are as follows:

- paranormal
- EVP
- instrumental transcommunication
- Raudive
- spirit
- ITC

Note that the use of abbreviations can produce irrelevant results, as they can also be used to represent company names, products, and other unrelated items.

Some recommend books about this fascinating subject, including material on EVP and EIP, are:

- *The Ghost of 29 Megacycles* (John Fuller)
- *Breakthrough* (Dr. Konstantin Raudive)
- *The Vertical Plane* (Ken Webster)
- *The Mediumship of the Tape Recorder* (D.J. Ellis)

- *Channeling: Investigations on Receiving Information from Paranormal Sources* (Jon Kilmo)
- *Encounters With the Paranormal* (Kendrick Frazier)
- *The World of Ted Serios* (Jules Eisenbud, M.D.)
- *Paranormal Powers—Secrets of the Unexplained* (Gary L. Blackwood, Daniel Cohen)
- *Beyond Light and Shadow* (Rolf H. Krauss)
- *America's Restless Ghosts* (Hans Holzer)
- *Beyond the Spectrum* (Cyril Permutt)

We also recommend the movie "The Grass Harp" to any reader who wants to gain a deeper understanding of this subject.

2.6 Combined Sound and Image Projects

It is an old maxim of mine that when you have excluded the impossible, whatever remains, however improbable, must be the truth.

Sir Arthur Conan Doyle (1859–1930)

Images and sounds can be combined in many paranormal experiments. Not only are the images picked up from the TV receiver suitable to experiments, but also static images that can be photographed with standard or Polaroid cameras.

Several experiments are suggested in the following pages, and some projects add special resources to the experiments for the experimenter who is looking for a different approach. There are also experiments involving sound and images in other paranormal areas such as ESP, psychometry, etc.

Sounds or images from the next dimension can be picked up, often when you least expect them, using the devices described next.

Project 16: Light-to-Sound Converter

When converting modulated light into sound using a sensor that is fast enough to respond to the changes, some surprising things can happen. For instance, if you point a sensor at a fluorescent lamp, you'll hear the 120 Hz signal of the modulated light, corresponding to the two instances in each power supply line cycle in which the voltage passes through zero volts. The gas inside the fluorescent lamp ionizes in each half-cycle of the power supply ac voltage and becomes inactive when the applied voltage between cycles crosses the zero point. If you point the same sensor to a white noise light source such as a small area of a TV screen, the sound produced is also a form of white noise.

Starting with this fact, a circuit that can convert modulated light into sound can be used in several experiments, not only when looking for paranormal voices but also to detect other paranormal phenomena such as ESP, psychometry, telekinesis, clairvoyance, etc. The circuit can also be used to receive modulated light from a Kerr cell or a laser in a wireless audio link.

The project described here is a simple light-to-sound converter that will allow you to "hear the light."

How It Works

The sensor is any phototransistor or photodiode. These devices have a light-sensitive surface that is accessible through a small window in the transparent plastic package. Light falling onto the sensor produces free current carriers that can be picked up by the circuit. If modulated light is the source, the current will have the same wave shape.

These sensors are very sensitive and can even "see" light wavelengths that our eyes can't (e.g., infrared and ultraviolet). They are also very fast (many times faster than our eyes), "seeing" changes in light amplitude that we can't detect. Some photodiodes can "see" variations in light amplitude at frequencies up to 50 MHz.

Applying an amplifier stage to the current, we can increase it to a level strong enough to drive a headphone or even a loudspeaker. The amplifier is formed by an LM386, an integrated circuit that can provide some hundreds of milliwatts to a small loudspeaker or a headphone when powered from four to six AA cells.

The gain of the circuit (about 200 times) is determined by C5. P1 acts as a sensitivity or volume control. An economical approach to building the sensor is the use of a common power transistor such as shown in Fig. 77.

By removing the metallic package to expose the chip, any silicon transistor can be converted into a phototransistor. Types such as the 2N3055 are sensitive enough to provide the circuit with a few milliamperes of current when illuminated. These components act as true photocells, converting light into electrical energy.

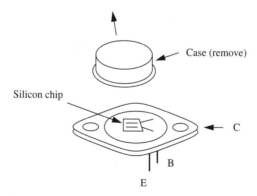

Figure 77 Exposing the silicon chip of a power transistor to convert it into a light sensor.

Assembly

The complete circuit of the light-to-sound converter is shown in Fig. 78. The circuit can be mounted as shown in Fig. 79.

Any phototransistor can be used as the sensor. It is important to install the sensor inside a small cardboard tube to pick up light from only one direction. Another resource for improving circuit sensitivity is the addition of a convergent lens in front of the sensor, placed as shown in Fig. 80. With the addition of the lens, the circuit becomes very directional and can pick up light from small, distant light sources.

Any low-impedance (4 to 100 Ω) headphone or even a 5 cm to 10 cm × 8 Ω loudspeaker can be used as final transducer to convert the signals into sound. You can point the sensor at a fluorescent lamp placed miles away from you and hear it using this circuit.

If a laser beam, placed at a distance of a few miles, is modulated by an amplifier, and if you use a microphone as an input to the amplifier, you can use this device to hear anything that is spoken into the microphone. The light beam will act as a carrier for the sounds.

Using the Circuit

Turn on the power supply and point the photosensor to a fluorescent lamp or to the screen of a TV receiver. The light signal converted into sound will be heard as a continuous tone. (In the case of TV, the sound you hear is the frame rate.)

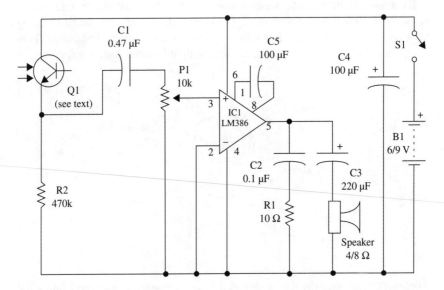

Figure 78 Light-to-sound converter.

Experimenting with Images 101

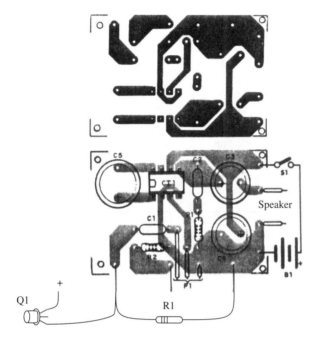

Figure 79 Printed circuit board for Project 16.

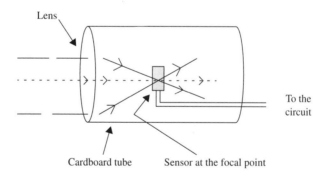

Figure 80 Increasing directivity and sensitivity with a lens.

Alter the value of R2 to change the circuit characteristics to accommodate the sensor that you are using. Depending on the sensor, this resistor can assume values between 100 and 2,200 kΩ.

You can experiment with picking up voices from a noisy image in the TV screen by plugging the output of this circuit into a tape recorder as shown in Fig. 81. This figure also shows how you can use the sensor to directly drive the input of any audio amplifier (including the ones described in this book) or the input of a tape recorder.

Suggestions

- Infrared or ultraviolet filters can be placed in front of the sensor to obtain a selective frequency response to light sources.
- Placing moving objects in front of the sensor, such as the helix of a fan, a stroboscopic effect can be added to the experiments.
- Colored filters, made with pieces of sheet cellophane, can be used in some experiments.
- Point the sensor toward the sky, and the natural light will produce some strange signals—at night.
- Thunderstorms can be heard if you point the device at the sky to pick up lightning flashes.

Parts List: Project 16

Semiconductors

IC–1	LM386 integrated circuit, audio amplifier
Q1	Any phototransistor (TIL71, TIL81, TIL411, etc.) or photodiode (see text)

Resistors

R1	10 Ω, 1/8 W, 5%—brown, black, black
R2	470 kΩ, 1/8 W, 5%—yellow, violet, yellow

Capacitors

C1	0.47 µF, ceramic or metal film
C2	0.1 µF, ceramic or metal film
C3	220 µF/12 WVDC, electrolytic
C4	100 µF/12 WVDC, electrolytic
C5	100 µF/12 WVDC, electrolytic
C6	10 µF/12 WVDC, electrolytic

Miscellaneous

P1	10 kΩ potentiometer
SPKR	4 to 100 Ω speaker or headphone
S1	SPST, toggle or slide switch
B1	6 to 9 V, four or six AA cells

Printed circuit board, cell holder, plastic box, convergent lens, cardboard tube, wires, solder, etc.

Experimenting with Images 103

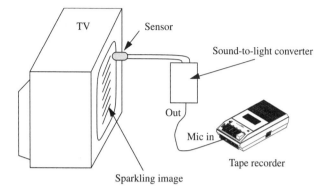

Figure 81 Converting random images into sound.

Project 17: Sound-to-Light Converter

This project converts the sound from a tape recorder, microphone, or even from a white noise source into light. The light will be modulated in amplitude with a response up to a few kilohertz. The circuit is very simple and also can be used in many experiments involving EIP and hypnosis.

How It Works

The circuit consists of a single transformer plugged into the output of any audio amplifier that has a power output in the range between 5 to 20 W. The transformer has a low-impedance winding that receives the audio signal, transforming it into a high-voltage signal. This high-voltage signal appears in the high-impedance winding where a fluorescent lamp is connected.

When applying an audio signal to the input of the amplifier, the signal is converted into high-voltage variations that make the fluorescent lamp flash at the same rate. If a white noise is used to drive the amplifier, the lamp will flash according to this source.

Even fluorescent lamps that no longer function when plugged into the ac power line can be used in this project. This is because the high-voltage spikes produced by the transformer can reach more than 400 V in some cases.

Observation

The frequency response of a fluorescent lamp is not high. Frequencies above a few kilohertz have no effect on the lamp. Because of this limitation, if you use a white noise source to modulate the lamp, the results will be less than excellent. Incandescent lamps are not recommended for this application, as their frequency response is even lower than that of a fluorescent (due to the filament inertia).

Assembly

Figure 82 is a diagram of the simple circuit. Figure 83 shows the actual appearance of the mounting. *Special care must be taken in connecting the wires to the lamp. The high voltage induced by the transformer can cause severe shock if the live parts of this circuit are touched.*

Any transformer with a primary winding rated from 117 to 240 V and a secondary winding rated from 9 to 12 V × 1 A or more can be used experimentally. The fluorescent lamp can be of any type with power rated to values between 5 and 40 W. Color fluorescent lamps or even ultraviolet fluorescent lamps can be used.

To plug the circuit into the output of an amplifier, use common wires with lengths up to 5 m. Long wires can cause losses in the signal, affecting the circuit performance.

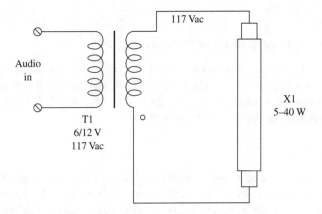

Figure 82 Simple sound-to-light converter.

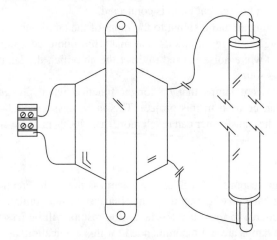

Figure 83 No chassis is needed to assemble this project.

Using the Circuit

The circuit is plugged to the speaker output of the amplifier. The audio source, a tape recorder with a tape of noise or other sounds, can be used for this purpose. Any white noise generator, including those described in this book, can also be used as an audio source.

Adjust the volume control of the amplifier to make the lamp flicker with the sound. If the lamp doesn't flicker, even when the amp is set to maximum volume, you probably aren't using a powerful enough amplifier, or the transformer is not suitable for the task.

Do not increase the volume beyond the point where the lamp starts to flicker. Doing so may overload the transformer or the lamp, or it may saturate the circuit.

If you use a powerful enough amplifier, you can drive the lamp and a loudspeaker at the same time. If you try this, be careful not to lower the impedance of the load to values below the minimum recommended. For instance, if your amplifier has a 2 Ω output, don't use a 2 Ω loudspeaker system with the lamp. Use a 4 Ω system with the lamp.

Experiments

Some interesting experiments can be made with this modulated light source. You can use the lamp powered from a white noise source to illuminate a wall and take photos from it. You can also use it to illuminate the wall and use your video camera and VCR to record images. The lamp powered from a white noise source or other source can also be used with the light-to-sound converter.

If you drive the amplifier using a special tape of gong sounds, recorded for transcendental meditation, the lamp will follow the tones with a visual output as well as audio.

Suggestions

- Use this circuit to illuminate a wall and try to take images from it using a photographic camera or a video camera.
- Use the circuit with the light-to-sound converter. Tape sounds from an illuminated wall.
- Replace the white lamp with a color or UV (ultraviolet) fluorescent lamp. Neon lamps can also be used. (A 100 kΩ series resistor is needed when using small neon lamps.)
- Mount two or more circuits to convert sound to light, and combine them when illuminating a wall or screen for your experiments. Plug one circuit into each output of a stereo amplifier. Use lamps of different colors for each channel.
- Appropriate sounds applied to the circuit (percussion sounds from a tape, for instance) can be used to induce sleep or to create a special mental state in transcendental meditation.

Parts List: Project 17

T1   Transformer, primary 6 to 12 V × 1A or 2 A (for amplifiers up to 20 W) or primary 6 to 12 V × 2 to 4 A (for amplifiers up to 100 W), secondary 117 Vac or 220/240 Vac.

X1   5 to 40 W, any fluorescent lamp (white, color, or ultraviolet) (see text)

Miscellaneous

Plastic box, wires, audio amplifier (5 to 100 W), etc.

Project 18: Brontophonic Sound

Brontophonic comes from the Greek word "brontos," meaning "thunder," and "phonos," meaning "sound." In other words, brontophonic means something like "sound of thunder."

The penetrating sound of the thunder sometimes creates the sensation that it comes from inside our head. This project is based on this fact, using a new concept of sound wherein the sensation of hearing is produced inside our brains. Although some researchers are trying to use this concept in audio (hi-fi) applications, including an author who published some articles about it many years ago, the brontophonic sound can also be used with other goals, including paranormal experiments as described here.

To give the reader an idea of what the brontophonic sound is, we begin by explaining the familiar physical phenomenon called *beat*.

Beat

In your secondary school or physics lab, you may have learned about tuning forks such as the one shown in Fig. 84. When a tuning fork is struck, it vibrates at its natural frequency, which is determined by its size, material, and shape. The sound produced this way is called "pure," as it has a sinusoidal wave shape and a specific constant frequency.

An important phenomenon occurs when two different tuning forks (tuning forks tuned to different frequencies) are set to vibrate at the same place and at the same time. If you listen carefully, you'll not only hear the two basic tones (for which they were designed), you may also hear a much lower tone and a much higher tone as well. This phenomenon occurs when the two tones are mixed inside your ear, or specifically in the eardrum membrane.

One of the additional frequencies heard is a *sum* frequency, and the other is a *difference* frequency. If one fork vibrates at 800 Hz and the other at 300 Hz, it is very possible that you will hear a lower frequency tone of 500 Hz (800 − 300 Hz) and a higher tone of 1,100 Hz (800 + 300 Hz).

This phenomenon, caused by interference between sound waves, is called *beat*. Many electronic devices such as radios, TV, and telecommunication circuits, use

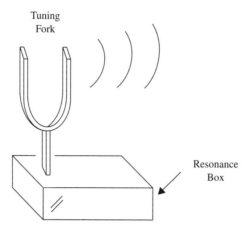

Figure 84 Common tuning fork.

the beat principle in their operation to lower the frequency of a signal to a value that can be used to drive the processing circuits.

If you examine how the two additional frequencies are produced (500 and 1,100 Hz), you'll find that the phenomenon occurs because each point of the eardrum simultaneously receives vibrations from two different sources and has to vibrate at two frequencies at once. The result is a combined movement of each point of the eardrum that produces the two additional frequencies.

The important point to notice in this phenomenon is that the two new frequencies are not produced by the sound sources but are created inside the ear, or in a microphone, at any observed point in space.

Let's now go a step further and transpose the same concepts to ultrasonic sources, one of them operating at 19,000 Hz and the other at 20,000 Hz. (Remember that we can't hear sounds at frequencies much above 18,000 Hz.)

The vibrations produced by either source separately can't be heard by anyone. But, when the two tones are mixed inside your ear, they produce two new tones. One of them, the sum tone of 39,000 Hz, is above our upper auditory limit, but the other, the difference tone of 1,000 Hz, is in the audible frequency range and can be heard.

The interesting part of this phenomenon is that, because the difference tone is produced inside your ear (or, more precisely, in your eardrum), you will have the strange sensation that the sound comes from inside your head, or that the sound comes from nowhere.

If the ultrasonic sounds are produced by high-power sources, the beat sound generated inside you will cause a certain discomfort, and exposure to vibrations of this kind can even cause panic or other psychological or physiological effects. (In some cases, it has been reported that very low-frequency beats can even cause dysentery!)

For the paranormal researcher, this kind of ultrasound source can be used in some fantastic experiments. An ambient can be filled from two ultrasonic sources and picked up by a microphone plugged into a recorder. The type of sound that can be heard when the tape is edited is a subject for discussion.

It is also interesting to see how this kind of vibration can affect or stimulate people with paranormal abilities or in ESP experiments.

How It Works

This project is formed with a pair of powerful ultrasonic modulated oscillators running in ranges between 18,000 and 25,000 Hz. A 4093 IC is used as basis of the project.

Each of the four NAND gates of the 4093 is wired as an oscillator. In the first block, we have IC1-a running in a very low frequency range (modulation circuit) given by R1 and C1. This circuit modulates (in frequency) the second oscillator, made with the IC1-b NAND gate.

This oscillator runs in a frequency range between 18,000 and 25,000 Hz, given by C3 and adjusted by P1. The frequency-modulated signal produced by this circuit is applied to a high-power output stage using a power FET.

As a load, the power FET uses a piezoelectric tweeter. Many common tweeters can reproduce sounds above our audible limit with good performance and therefore can be used as ultrasonic transducers. Common piezoelectric tweeters can produce several watts of ultrasonic sounds in the range up to 25,000 Hz. Pay attention to the characteristic of the tweeter you use, choosing one that can go up to 25,000 Hz or more.

The other low-frequency oscillator has its frequency determined by C5 and R7, and it modulates the fourth oscillator, made with IC1-d. This oscillator operates in the ultrasonic range and has its frequency adjusted by P2. The signal from the second ultrasonic oscillator also drives a power output stage with a power FET, and a second tweeter is used as transducer.

The circuit can be powered from 6 to 12 V sources and can fill an ambient with several watts of ultrasonic vibrations as described above. It is only necessary to place the tweeters a few meters apart to fill one ambient with different ultrasonic signals, producing the strange effects described in this section.

Warning!

Mammals such as dogs, cats, rats, et al. can hear not only the audible beat produced by this device, but also the ultrasonic sounds. These animals are made very uncomfortable by the sounds generated in this experiment. Don't use this equipment if they are present.

Some low-frequency beats that can occur with this device, in the range of alpha, beta, and theta brain waves, can induce epileptic seizures. Do not use this device in experiments that would require people to remain in the ultrasonic field during prolonged sessions! Avoid the use of this device near anyone who has a history of epilepsy.

The very low-frequency beats can create additional physiological/psychological effects in humans and, in some cases, may even cause dysentery!

Assembly

Figure 85 shows the schematic diagram of the brontophonic oscillator. The printed circuit board for this project is shown in Fig. 86.

Any P-channel power FET, rated to 4 A or more of drain current and voltages of 200 V or more, can be used in this project. Types of the IRF series, common in switched-mode power supplies such as found in computers, are cheap and suitable for this project.

The transistors must be mounted on heatsinks. The heatsinks are pieces of metal, 5 × 8 cm, bent to form a "U" and screwed to the transistors.

The small tweeters, as shown in the figures, with power ratings of 50 W or more are suitable and can be plugged to the circuit with wires from 40 cm to 2 m in length. The tweeters must be placed as far apart as possible for best results.

A power supply suitable for this circuit is shown in Fig. 87. The transformer has a primary winding rated to the ac power line voltage and a secondary rated to 6 to 9 V and 2 A of current.

The circuit can also be powered from four to six D cells. As the current drain is high, the continuous operation time when using cells is not long. A plastic or wooden box can be used to house the circuit.

Although, in the original project, we recommend trimmer potentiometers, the reader is free to modify the project to use common potentiometers placed in the front panel. This will allow the researcher to have additional control over the generated signal.

Using the Circuit

Place the two tweeters separated from each other by a distance of at least 2 m. Turn on the power supply.

Adjust P1 and P2 until you get the strange sensation of modulated sound being produced inside your brain. Do not expose yourself over a long period to this sound, as it can produce panic and other unpredictable dangerous effects.

Warning!

If using the equipment in experiments involving humans, take care and consult a specialist to avoid problems. Ultrasonic sounds and beats are dangerous and can cause problems that include the possible inauguration of an epileptic attack!

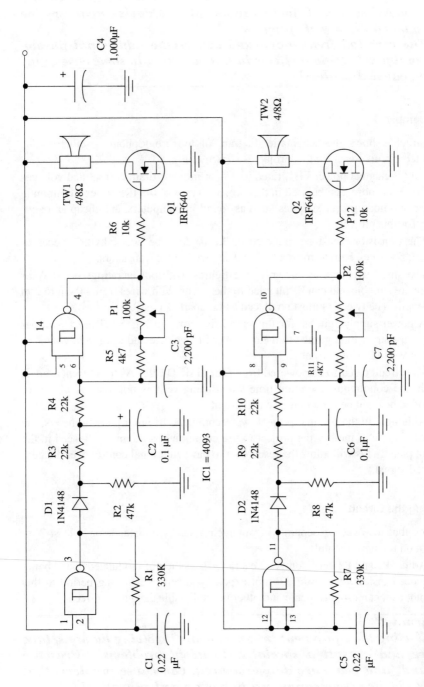

Figure 85 Brontophonic sound generator.

Experimenting with Images 111

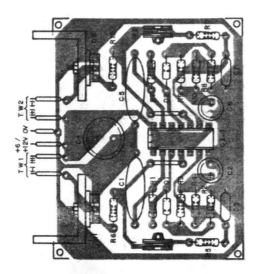

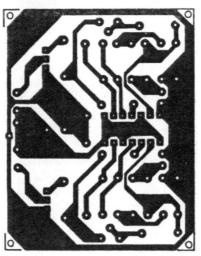

Figure 86 Printed circuit board for Project 18.

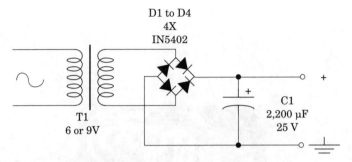

Figure 87 Power supply for Project 18.

When attempting to pick up voices, do not remain in an ambient that is filled with the signals produced by this device.

Observation

The device can be used in alarm systems, as it can drive an intruder into a state of total discontrol and force him to leave before accomplishing his intent.

Suggestions

- A volume control can be added as shown in Fig. 88. With this control, you can perform the experiments in the presence of humans without the discomfort of a high-power ultrasonic source.

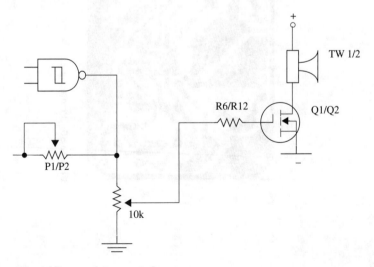

Figure 88 Adding a volume control.

- Only one oscillator can be used in experiments involving pure ultrasound. You can add a switch in series with each tweeter to turn it on and off whenever you want.
- You can add a switch to disable one or the other oscillator to perform experiments with only one ultrasonic source.
- A special low-impedance ultrasonic transducer can be tested with this circuit. Depending on the transducer used, the frequency can be increased to more than 30,000 Hz.
- The high-power ultrasonic signal produced by this circuit can scare birds from your property. You can conduct further experiments in this area.
- In complex experiments, you can use more than one of these circuits to fill an ambient with a variety of ultrasonic sounds.

Parts List: Project 18

IC-1	4093, 4 NAND Schmitt Gates, CMOS integrated circuit
Q1, Q2	IRF640, IRF720, IRF620 or equivalent. Any power FET (field effect transistor) (see text)
D1, D2	1N4148, 1N914 or equivalent, silicon general-purpose diodes

Resistors

R1, R7	330 kΩ, 1/8 W, 5%—orange, orange, yellow
R2, R8	47 kΩ, 1/8 W, 5%—yellow, violet, orange
R3, R4, R9, R10	22 kΩ, 1/8 W, 5%—red, red, orange
R5, R11	4.7 kΩ, 1/8 W, 5%—yellow, violet, orange
R6, R12	10 kΩ, 1/8 W, 5%—brown, black, orange

Capacitors

C1, C5	0.22 µF, ceramic or metal film
C2, C6	0.1 µF, ceramic or metal film
C3, C7	2,200 pF, ceramic
C4	1,000 µF/16 WVDC, electrolytic

Miscellaneous

P1, P2	100 kΩ potentiometers or trimmer potentiometers (see text)
TW1, TW2	Piezoelectric tweeters (see text)

Printed circuit board, power supply, plastic or wooden box, heatsinks, knobs for P1 and P2, wires, solder, etc.

Project 19: Ultrasonic Sound Converter (Hearing the Unhearable)

Interesting experiments involving inaudible sounds can be performed with this circuit. This project describes a device that converts inaudible sounds (ultrasonic sounds) into audible sounds. The researcher can use it to examine inaudible signals recorded on tape or picked up by radios or other devices, or to hear sounds picked up in an ambient by a transducer.

Our circuit is a simple version of a digital frequency divider. The ultrasonic sounds that have been taped or picked up by a transducer are applied to the input of the circuit, which divides their frequency by a number chosen by the operator. In this manner, in the output of the circuit, the signal will drop to a frequency that is within the audible range. For instance, if a 30 kHz signal (ultrasonic) is applied to the circuit, which is programmed to divide its frequency by 6, in the output we hear will be a 5 kHz signal (in the audible range).

Some limitations are evident in this circuit due its simplicity. One of them is that the circuit operates with signal spikes, as it is digital, and the output is a rectangular signal. This means that the original wave shape is not preserved. The "thing" we are going to hear from the output is not a true conversion of the original sound, but something that corresponds to it.

The circuit may be upgraded in several ways, such as the addition of filters to get improved results, or by some form of signal processing. The basic idea is intended to stimulate the reader's imagination toward creating new projects.

Another kind of upgrade to be considered by the readers is a wave shape conditioner, which can be made with a Schmitt trigger (a 4093, for instance) to convert fast changes of signal level into digital signals that could be processed more easily by the frequency divider.

How It Works

The circuit is formed basically by a CMOS integrated circuit acting as a frequency divider and a low-power audio amplifier driving a small loudspeaker or a headphone.

The divider is a CMOS 4020 IC, a 14-stage binary divider formed by a chain of flip-flops, each dividing the frequency applied to its input by two. Another option for the frequency divider is the 4040 CMOS IC.

We use outputs Q4, Q6, Q9, Q12, and Q14, which are preset to perform divisions by 16, 64, 512, 4,096, and 16,384. If we apply a 32 kHz signal to the circuit and take it from Q4 (divided by 16), the signal will be lowered to 2 kHz, falling in the audible range.

The reader is also free to modify the original project using other output of the CMOS 4020. These offer division by 2, 4, 8, 16, 32,..., 16,384.

If a 4 MHz signal is applied to the input and the circuit is programmed to divide by 4,096 (Q14), the output will be a 4 kHz signal (in the audible range). We must remind the reader that 4 MHz is just near the upper limit of input signals for the 4020 when powered from a 12 V supply. If the circuit is powered from four AA cells (6 V), as in this application, this upper limit will fall to about 2 MHz.

The circuit can be powered from AA cells or a 9 V battery It also can be used in interesting experiments involving ultrasonic sounds and even high-frequency signals, not only when editing tapes or picking up paranormal sounds but in many other applications as well.

Assembly

The complete schematic diagram of the ultrasonic converter is shown in Fig. 89. A printed circuit board for this project is suggested by Fig. 90.

To program the frequency division in this circuit, the reader can use a rotary switch or a terminal strip with screen. The interconnection between the screws will determine the division rate of the circuit, as shown in Fig. 91.

Diodes D1 and D2 are used to protect the input of the circuit against spikes that can cause damage to the IC. We must remember that spikes with voltages above the power supply voltage can be dangerous to the IC. So, some care must be taken when adjusting the level of the input signal by P1. This means that P1 is not only the sensitivity control but also the limit control to the input signals.

P2 is used to adjust the output volume in the loudspeaker or headphone. For the input to the circuit, we can use an RCA jack or two wires with alligator clips.

All the components can be housed in a small box whose dimensions are basically determined by the size of the loudspeaker (if used—you can replace the speaker with a headphone).

Using the Circuit

When using the circuit with signals from a source such as a tape recorder or an amplifier, apply the signal to the input of the circuit starting from the lower volume (P1 at the medium). Then, gradually increase the volume of the signal source until you hear something in the output. Don't go far from this point, or you may saturate the circuit or damage the IC.

Using the Circuit with Tranducers

Many tweeters (high-frequency loudspeakers) can be used as ultrasonic transducers to obtain a reasonable gain (sensitivity) in frequencies up to 25 or 30 kHz. It is interesting to note the fact that a tweeter can be used both as a source of ultrasonic sounds and as a receiver (microphone) for them. You only have to open it up and directly access the transducer as suggested in Project 8. A one-transistor amplifier stage for this transducer is shown in Fig. 92.

Any general-purpose silicon transistor can be used in this circuit, such as the BC548, 2N2222, etc. The power supply is the same as for the main circuit. As the current drain of this stage is very low, it doesn't reduce battery life appreciably.

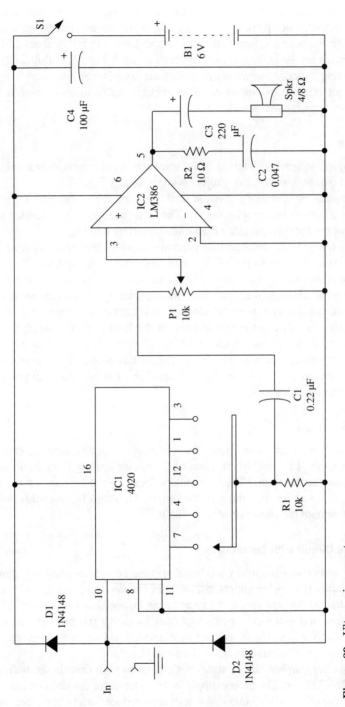

Figure 89 Ultrasonic converter.

Experimenting with Images 117

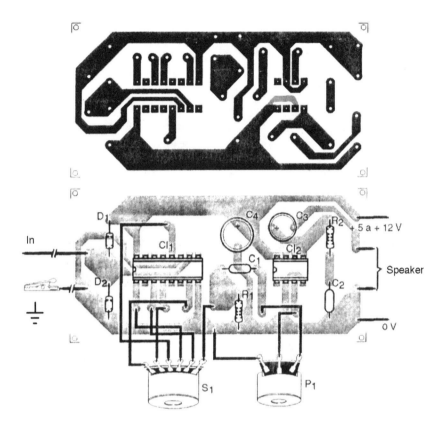

Figure 90 Printed circuit board for Project 19.

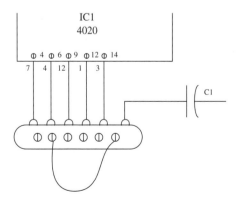

Figure 91 Dividing the input frequency by 64.

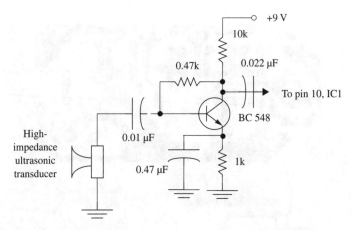

Figure 92 Transducer preamplifier for low-impedance tweeter used as sensor.

Sounds between 15 kHz and 25 kHz or more can be heard in the headphones or loudspeaker using this project. If it is used to find voices by lowering the frequency of a recorded signal, plug the earphone into the input of the circuit and adjust the volume of the output until you hear the desired signal in the output. Find the correct value of frequency division by performing the recommended experiments.

Suggestions

- This circuit can be used to pick up voices in an ambient filled with ultrasound. Its output should be plugged into the input of a tape recorder.
- A capacitor with values between 10 nF and 1 µF can be placed between pin 3 of IC2 and the negative power supply rail to filter the sound.
- The filters described in the projects for EVP can be used to process the sounds picked up with this circuit.
- If a high-frequency signal is applied directly to the input of the circuit, a detector diode must be wired as shown in Fig. 93. The circuit can work with signals up to 2 MHz when powered from a 6 V power supply.

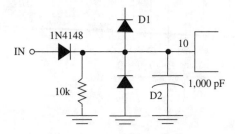

Figure 93 Working with high-frequency signals.

- Directivity can be added to the microphone (ultrasonic transducer) if, for example, it is placed inside a metallic parabolic reflector. (The microphone must be placed in the focus of the parabolic reflector.) Diameters between 40 cm and 1 m are suggested for the parabolic reflector.

Parts List: Project 19

Semiconductors

IC1	4020 CMOS integrated circuit, 14-stage binary dividers
IC2	LM386 low-power audio amplifier integrated circuit
D1, D2	1N4148 or equivalent general-purpose silicon diodes

Resistors

R1	10 kΩ, 1/8 W, 5%—brown, black, orange
R2	10 Ω, 1/8 W, 5%—brown, black, black

Capacitors

C1	0.22 µF, ceramic or metal film
C2	0.047 µF, ceramic or metal film
C3	220 µF/12 WVDC, electrolytic
C4	100 µF/12 WVDC, electrolytic

Miscellaneous

P1	10 kΩ potentiometer
S1	SPST—toggle or slide switch
B1	6 V, four AA cells
SPKR	4 or 8 Ω, 5 to 10 cm, loudspeaker (or headphone) (see text)

Printed circuit board, input and output jacks (optional—see text), cell holder, plastic box, wires, solder, etc.

Project 20: Laser Image Generator

A laser pointer or a neon laser can be used as light source in some experiments involving paranormal images. Other applications for this circuit include experiments in transcendental meditation, ESP, etc.

The basic idea is the use of a modulator to make the light beam change its direction according to signals from a sound source and then to draw figures on a screen, as shown in Fig. 94.

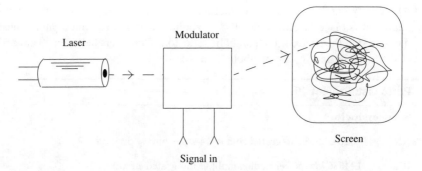

Figure 94 The laser beam changes direction according to the input signal.

Using any signal or noise source in the modulation process, the figure produced in the screen can reveal strange forms when photographed. Of course, the circuit has other applications, such as for light shows, and it produces interesting visual effects when the modulation circuit is plugged to the output of an audio amplifier. Upgrading the circuit with a two-axis mirror control, the circuit can be plugged into the output of a processor to draw images on a screen.

How It Works

A small mirror is affixed to a loudspeaker as shown in Fig. 95. When the loudspeaker's cone moves forward and back, the mirror changes its position, altering the reflection angle of the laser beam. As the cone's movement is fast, the reflected beam will draw random figures on a projection screen. If a photographic camera is placed in front of the screen with the shutter open (time exposure), the figures produced by the beam can be recorded on film.

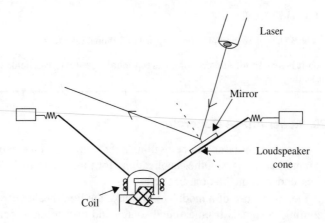

Figure 95 A small mirror placed on the cone of a loudspeaker.

Assembly

Using glue, affix a small mirror to the cone of a 4 to 6 inch loudspeaker. Super glue or an epoxy is recommended.

The loudspeaker is connected to the output of any low-power amplifier (0.5 to 5 W). Don't use high-power amplifiers, which can damage the loudspeaker if you turn the volume up too high.

In normal operation, you only need several milliwatts to generate an image on a wall. The size of the image depends on the distance between the mirror and the wall.

The laser must be placed in a position that allows the beam to be pointed at the mirror and reflected to the wall or screen. A tripod or other stand can be used for this task.

Warning: Laser radiation is dangerous. Avoid direct exposure to the beam.

Using the Circuit

Point the laser at the mirror so that it produces a light point on the screen. The screen can be white cardboard or any white wall.

Plug the audio source into the input of the amplifier. Any of our noise sources can be used as an audio source in experiments involving EVP and EIP.

Adjust the audio amplifier volume control, starting from the minimum to produce an image of the proper size. (The "proper size" depends on the distance between the screen and the camera you will use to record the images.) To make experiments with a photographic camera, open the lens for a period of 15 to 40 seconds, registering the images in the interval. Other light sources in the ambient must be extinguished to avoid overexposing the film.

Remember that the loudspeaker doesn't have much high-frequency response. This limitation is a factor when working with high-frequency signals and also with white noise.

Suggestions

- The mirror can be moved by two loudspeakers as shown in Fig. 96, combining sounds from two sources. In this experiment, the mirror will produce two-axis movements that are determined by the signals applied to the loudspeakers.
- You can use other light sources in the experiments, such as colored spotlights, flashlights, etc.
- Audio filters can be added in series with the loudspeaker (see experiments with EVP) to allow only specific frequency ranges to affect the experiments.

Observation

This circuit can be used for light shows in clubs or for other public events. It will produce dancing images when the loudspeaker is plugged into the output of the

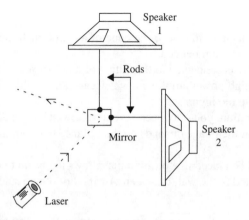

Figure 96 Differential or two-axis modulator.

public address (or other) amplifier. If the loudspeaker is connected to a high-power amplifier, use a resistor (100 Ω to 1 kΩ) to reduce the applied audio power to safe levels.

Parts List: Project 20	
SPKR	4 or 8 Ω × 4 or 6 inch (10 or 15 cm) loudspeaker
X1	Mirror (2 × 2 cm)
M	Audio amplifier (see text)
L	Laser source (laser pointer or neon laser)
Miscellaneous	
Wires, video or photographic camera, screen, etc.	

Project 21: Magnetic Field Generator

The influence of magnetic or electric fields in some paranormal experiments has been studied by many researchers. Placing magnets near the subjects or near a microphone when trying to pick up images is one experiment that has been performed many times. The presence of a magnet can also enhance the reception of images from walls or other places. Magnets are also used in experiments involving other paranormal phenomena such as ESP, clairvoyance, telepathy, etc.

The kind of influence that a magnetic field can have on these experiments is speculative and not included in this discussion. But we should note that the common magnet has an important limitation: it produces a constant magnet field that, in some cases, is not strong enough for our purposes.

The magnetic field generator presented in this project can offer something more to the paranormal researcher. It produces a variable low-frequency magnetic field.

Several paranormal experiments can be performed with the aid of this project to explore the following subjects:

- The influence of an alternating magnetic field when picking up sounds from paranormal sources
- The influence of an alternating magnetic field when registering paranormal images
- How ESP experiments are influenced by magnetic fields
- The influence of the magnetic field in experiments involving Kirlian photography, telepathy, precognition, and psychokinesis/telekinesis
- The influence of magnetic fields on plant growth and other physiological reactions of plants (Backster effect)

How It Works

The circuit is a low-frequency oscillator with a versatile configuration that can be used with several purposes, as the reader will see in this book. The low-frequency signal produced by the oscillator can be used to drive a coil, producing a magnetic field, or to drive a transformer, producing high voltage in Kirlian experiments (as described in Section 2.7).

The operational principle is simple: when a current flows through a coil, a magnetic field is produced as shown in Fig. 97. The direction of the magnetic field lines depends on the current direction. As this circuit produces direct pulsed current, the direction is determined by the way the coil is connected. If required for a particular experiment, a switch can be added to invert the direction of the magnetic field.

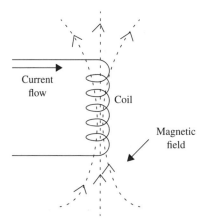

Figure 97 Magnetic field in a solenoid.

124   Part 2

The frequency of the pulse current is also the frequency of the field and can vary in a range from several hertz up to 10,000 Hz. Capacitor C1 determines the frequency range and can be altered as suggested below. R1, R2, and P1 determine the frequency with C1. The duty cycle is determined by R1, which can be chosen in the range between 4.7 and 47 kΩ. R2 and P1 determine the pulse rate or frequency. With the recommended values for C1, the frequency is in the range between several hundreds of hertz and several kilohertz. By reducing the value of C1, the frequency can increase to values up to 1 MHz. The pulsed signal is applied to a power FET that drives the coil, producing large currents.

The circuit can be powered from supplies ranging from 6 to 12 V. The output power or intensity of the magnetic field depends on this voltage. You can use any of the power supplies recommended for other projects in this book as long as they have output voltages between 9 and 12 V and currents in the range between 500 mA and 2 A. Four D cells can be used to power the circuit, but only for short-term experiments, as the current drain is significant.

Assembly

The complete schematic diagram of the magnetic field generator is shown in Fig. 98. A printed circuit board is used to support the basic components, as shown in Fig. 99.

Any P-channel power FET with drain currents of 4 A or more can be used in this project. Types such as the IRF640, and many others in the "IRF" series, can

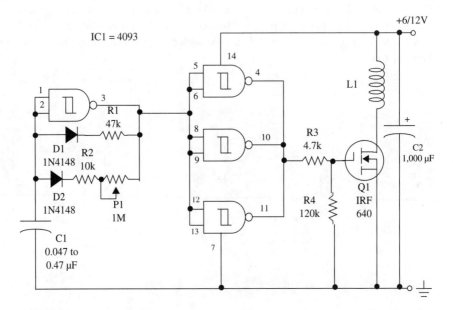

Figure 98 Magnetic field generator.

Experimenting with Images 125

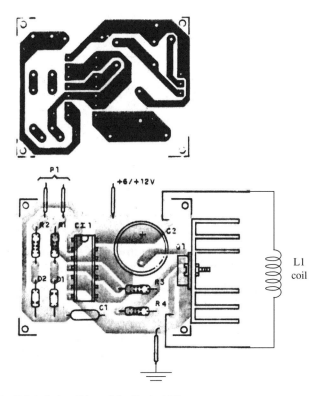

Figure 99 Printed circuit board for Project 21.

be used. Types such as the IRF630, IRF640, and IRF720 are suitable for this project. An alternative is the use of an NPN Darlington power transistor with collector currents rated to 4 A or more.

The power FET needs a heatsink, made with a piece of metal bent to form a "U." The coil depends on the application the reader has in mind. To apply the magnetic fields to small subjects (such as plants, a person's hands, or even the head), the coil can be made as shown in Fig. 100.

On a cardboard or wooden form, 20 to 50 turns of enameled wire (28 to 32 AWG) are wound and plugged to the circuit using a 1 to 2 m common, plastic-covered wire.

Warning: Some studies have indicated that prolonged exposure to low-frequency magnetic fields can cause diseases such as cancer and leukemia in humans. But, at the same time, researchers also have discovered that some low-frequency magnetic fields can help to restore broken bones and, in some cases, accelerate the recovery of injuries in some people. If the reader intends to use this device to expose persons or living

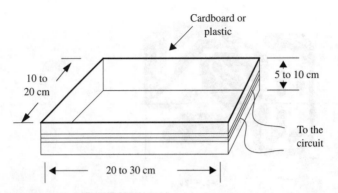

Figure 100 The coil.

beings, it is advisable to proceed with great prudence. Don't expose yourself or others to the magnetic field for more than a few minutes when making any experiment.

If the circuit is powered from the ac power line, the power supply shown in Fig. 101 can be used. The transformer has a primary winding rated to 117 Vac (or according your ac power line voltage) and a secondary rated to 6 V to 9 V × 1 A. The diodes are 1N4002 or equivalents, and the electrolytic capacitor is rated to 12 WVDC. Although the transformer isolates the circuit from the ac power line voltage, it is important to take precautions to isolate all high-voltage components to avoid receiving shocks from any part of the circuit.

Testing and Using the Circuit

Turn on the circuit and, inside the coil, place a MW receiver tuned to a free point between 550 and 1600 kHz. You'll hear a tone if the circuit is operating and a magnetic field is being produced.

The signals are picked up by the receiver at higher frequencies, as the circuit produces a rectangular current that is rich in harmonics extending up to the shortwave band.

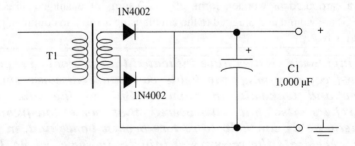

Figure 101 Power supply for Project 21.

Experimenting with Images 127

To apply the magnetic field to any object or sample, simply place it inside the coil. Adjust the frequency according to what is desired for the experiment.

Suggestions

- Living beings such as plants and insects can be placed inside the coil for a predetermined time interval before beginning the experiments. The influence of the magnetic field can be studied by comparing samples that have and have not been subjected to the magnetic field.
- The magnetic field can be used in ambients where voices and images are being picked up in paranormal experiments to see how it can influence the results.
- D1, D2, and R1 can be removed. The circuit then will produce 50% duty cycle signals.
- Replace the transistor with a Darlington bipolar NPN transistor such as the TIP140, TIP142, or TIP142.

Parts List: Project 21

Semiconductors

IC–1	4093 CMOS integrated circuit
Q1	IRF640 or any power FET (see text)
D1, D2	1N4148 or equivalent—any general-purpose silicon diode

Resistors

R1	47 kΩ, 1/8 W, 5%—yellow, violet, orange
R2	10 kΩ, 1/8 W, 5%—brown, black, orange
R3	4,700 Ω, 1/8 W, 5%—yellow, violet, red
R4	120 kΩ, 1/8 W, 5%—brown, red, yellow

Capacitors

C1	0.047 to 0.47 µF, ceramic or metal film
C2	1,000 µF, 16 WVDC, electrolytic

Miscellaneous

P1	1 MΩ potentiometer
L1	Coil (see text)

Printed circuit board, plastic box, wires, 28 to 23 AWG enameled wire for the coil, solder, etc.

Materials for the Power Supply

D1, D2	1N4002 silicon rectifier diodes
T1	Transformer—primary according the ac power line and secondary rated to 6 to 9 V CT, 1A

Power cord, 1 A fuse, wires, solder, etc.

2.7 Kirlian Photography

> *Ce qui a t cru par tous, et toujours, et partout, a toutes les chances d'tre faux. (What is believed by all, everytime and everywhere, has all the chances of being false.)*
>
> Paul Valry (1871–1945)

2.7.1 History: The Medical Aspect

Kirlian photography or *bioelectrography* is the name of the process involving the registration and observation of star-like patterns produced by high-voltage electrical discharge in living beings and objects. The patterns are associated with the "aura" of living beings.

The roots of the Kirlian photography are in observations of patterns produced in resin dust by high-voltage discharges. Early observations were made by the German scientist Georg Christoph Lichtenberg in 1777. He was the first person to observe a corona discharge from a human hand.

After the invention of photography, the Czech physicist Bartholomew Navratil, and the Russian-Polish electrotherapist, engineer, and physician Yakov Narkiewicz-Yodko, registered the first images of electric discharges from living beings and objects and started with a systematic observation of the effect. It was Navratil who first used the term *electrography* to describe the effect.

However, the modern history of the Kirlian Photography began with Semyon Davidovitch Kirlian (1900–1980). He is said to be the first person to build a Kirlian camera (in 1939) when, by chance, he rediscovered the effect that now bears his name. Kirlian made the first experiments with his camera in Krasnodar, in the former Soviet Union.

Numerous experiments were conducted by a student, Viktor G. Adamenko, along with Kirlian and his wife. Adamenko presented the first doctoral thesis on the subject at the Minsk Polytechnic Institute (Belorussia) and described the physical mechanism of image formation by the corona discharge process. Viktor based his thesis on the cold emission of electrons from the specimen due to high-voltage fields.

But another researcher, Victor M. Inyushin (a professor of biophysics at Kazakh State University, Alma Ata, Kazakhstan), along with Wlodzimierz Sedlak, developed the hypothesis of *bioplasma* to explain the Kirlian Effect in bioelectrography.

The Kirlian Effect became known in the West after the publication of two books, *Psychic Discoveries Behind The Iron Curtain,* by Ostrander and Schroeder (1970), and *The Kirlian Aura,* by Kroppner and Rubin (1974).

The first scientist to work on the Kirlian Effect in U.S.A. was Thelma Moss, a medical psychologist of the Neuropsychiatric Institute at the University of California at Los Angeles.

Another important name associated with contributions to the field is Gary K. Pook, professor of operations research and man-machine systems at the U.S. Navy Postgraduate School at Monterey, California. He was the first to introduce the bioelectrographic motion picture method outside the Soviet Union. He used a light intensifier for improving brightness, thereby revealing a dynamic aspect of the discharge process that is invisible even with long exposure times using still photography.

After 1970, based on information found in specialized magazines, the author designed for Dr. Max Berezovsky, M.D. (Brazil), a Kirlian machine. Using the apparatus, Dr. Berezovsky started with experiments in bioelectrography. Thousands of photos of the "aura" of objects and living things, such as plant leaves, have been made by that researcher.

An important document in this field was published in 1976 and 1978 by the U.S. Department of Defense. This document told about a six-year experiment conducted by a team headed by William Eidson, professor of physics at Drexel University in Philadelphia. The central point in this work was the discussion, among other things, of the typical instability of related equipment and the wide range of parameters that have to be controlled for best operation. One of the most important conclusions of this work was that electrography can produce images of the electrical parameters of a specimen in real time, making possible a mapping tool for the unexplained fields that surround living beings. These fields, to a certain degree, have the ability to modulate the characteristics of the surrounding space.

Another concept was developed by the German naturopathic doctor Peter Mandel in 1986, and it is now widely used as a method for interpreting the Kirlian image for medical diagnosis. In this application, the corona of the fingertips is related to their function as endpoints of the meridians in traditional Chinese medicine and German neo-acupuncture.

Bioelectrography or Kirlian photography is considered to be an inexpensive and quick method for the investigation of biological objects, based on the interpretation of an image obtained during their exposure to a high-voltage field. However, some problems have been found by the researchers, including

- Lack of systematic clinical studies with adequate sample numbers for statistical validity
- No theoretical basis for the mechanism of interaction between the corona discharge and the samples
- Difficulties when reproducing the results due to the lack of technical standards
- Lack of standard research methods

For this reason, in a workshop at the University of Aarhus (Denmark), the European Research and Standardization Group (ERSG) of the International Union of Medical and Applied Bio-Electrography (IUMAB) has been formed to discuss Kirlian photography as a diagnostic tool to be used in medicine.

2.7.2 History: The Paranormal Side

The presence of a luminous emission involving living beings, and even objects, soon was linked to an "aura" or a "bioenergy field" associated with the spirit. Many paranormal researchers began their investigations with efforts to associate the "aura" to paranormal manifestations or to a person's personality, mood, or state of health.

The color of the aura is also studied by many researchers as a way to detect the state of a person's health or even his mood. *Colortheraphy* is the term for the parascience that associates the color of an aura to the health and mood of a person.

One of the first experimenters in the field was the Brazilian Padre (Father) Landell de Moura, who, as a subordinate of the Roman Catholic Church, couldn't discuss that kind of invention. They say that the plans for his camera and the actual device were confiscated and have disappeared. Readers interested in the work of Landell de Moura can find much information about it on the Internet. One of the sites that of my friend Luiz Netto,[*] who wrote many articles about Landell de Moura in the magazines for which this book's author is the technical director. (See the "Chronology of the Transcommunication" in this book for more information.)

Even now, we can say that the Kirlian photography can be used in paranormal research. The images of the "aura" are powerful tools for the indication of paranormal phenomena, and many experiments can be performed with a Kirlian machine.

Kirlian Photography Defined. Kirlian photography (also known as *bioelectrography* or *electrophotography*) is a means of taking pictures of objects and living beings to reveal invisible patterns. These patterns are associated with the health or paranormal states of the objects to be photographed.

Paranormal researchers believe that the patterns registered in the plates by this process are associated to "forces and fields" that are present in all living beings, and also objects, and that these patterns can serve as instruments for diagnosis and detection of paranormal properties. In medicine, researchers believe that the patterns can reveal diseases or confirm normal, healthy states in living beings.

How Kirlian Photos Are Made. Kirlian photography is based on the well known physical phenomenon called *corona effect*. When an electrode is charged with a very high voltage (typically over 5,000 V), the charges tend to escape from the electrode by ionizing the adjacent air. The ionized air becomes conductive, producing a cold light radiation in the visible spectrum as well as in the ultraviolet and infrared spectra.

---

[*] A web site dedicated to Father Landell de Moura is at http://www.geocities.com/Athens/Olympus/4133/english.htm.

If a body is charged from a high-voltage source, the charges escaping into the air cause light to emanate from it, which gives it the qualities of an *aura*. Figure 102 shows the aura or corona effect produced when the hand of a person is placed on the electrode of a Kirlian machine. Kirlian observed the aura while working with high-voltage circuits of TV receivers and created an apparatus to photograph it.

The basic apparatus is formed by a high-voltage source and a plate with a metallic electrode where the high voltage is applied (4,000 to 50,000 V in the circuits commonly used in Kirlian cameras). A piece of glass is used to isolate the high voltage from the object or living being to be photographed.

The high voltage doesn't create a flux of charges that can cause electrical shock if someone touches the glass plate, but a high enough voltage is induced that the object, such as a finger, can radiate its own flux of charges into the air, creating an aura such as the one shown in Fig. 103.

If any conductive object is placed on the glass, an aura is produced by the induction of the high-voltage charge. The aura will cause the object to draw patterns corresponding to the parts in which the charge flux is higher or lower.

The flux of charges emitted from the object into the air depends on several factors, such as the existence of areas that are more highly conductive than others. All of this means that the pattern of the aura changes according the nature of the object placed on the electrode.

If a sensitive paper or film is placed between the object and the glass electrode, such as photographic film or even thermographic fax paper, the flux of charges can register an image on it. When revealed, the photo or figure on the paper will correspond to the electric flux of charges from the object during a particular time; i.e., the *aura,* as paranormal researchers call it.

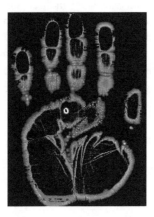

Figure 102 Aura of the human hand taken by a Kirlian camera. Analysis of the color version can reveal health problems in patients, according to paranormal researchers and many medical doctors. *Source:* photo courtesy of Parascope (color photo can be viewed at www.parascope.com).

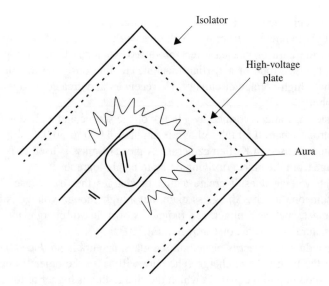

Figure 103 Aura of a fingertip.

This means that a basic Kirlian camera, represented in Fig. 104, is formed by (a) a high-voltage source, (b) a metal electrode, (c) a piece of glass to isolate the sample from the high-voltage source, and (d) a piece of film or sensitive paper to register the images. The object to be photographed is placed between the electrode and the film.

If the experiments are made in a dark place, the researcher can see the aura around the object placed on the electrode. This aura appears as a color luminescence involving the object and extensions in directions that depend on the nature of the sample. Blue, red, and yellow are the predominant colors in the aura of common living things, such as leaves, fingertips, hands, etc.

Many advanced Kirlian machines include video cameras to transfer the image to a computer or a TV. The images captured by the cameras can be manipulated by any image editing software.

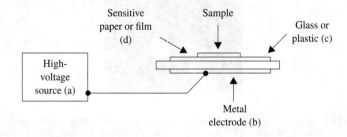

Figure 104 The basic Kirlian machine.

For practical reasons, some technical details must be observed when building a Kirlian machine. For instance, best results are found when the high-voltage source operates in the medium frequency range (between 2 and 10 kHz). There are many simple circuits that can produce these voltages from the ac power line, and we will discuss some of them in the following pages.

Caution!

The most important fact to remember when working with these circuits is that the high voltages are dangerous, and much care must be taken when working with any high-voltage devices.

To avoid shock hazards, the high-voltage sources are current limited. Even so, only readers who have proper knowledge and experience with high-voltage electronics should handle this equipment. The high voltages used in Kirlian cameras are dangerous and can be deadly under certain conditions. Although great care has been taken in the description of these devices, the author wishes to alert readers to the dangers of using them without proper installation and due caution.

What the Aura Represents in Paranormal Research. Kirlian photography is a new subject of investigation, so there are many theories and much speculation about the meaning of the image shapes and colors captured in the photos. Medical doctors are trying to define standards to associate the color and shape of the aura to the health of a person. Others are using the auras to diagnose the mental state of people and even to determine personality traits from the colors and shapes detected in the aura. Of course the images are real, but there is also misinterpretation resulting from the natural human tendency to extrapolate when interpreting results.

This first explanation for the patterns in the photos is related to the existence of *life energies* or *bioenergies* emanating from living beings. But, as in the case of EVP and EIP experiments, we call the reader's attention to possible misunderstandings in the use of these terms when trying to explain natural phenomena or anything that is unknown.

For readers who want to perform their own experiments with Kirlian photography, some facts about researchers' terminology on this subjects are important:

- Researchers found that the various colors picked up by the Kirlian photos have important meanings. They suggest that normal, healthy living beings "emanate" clear, white coronal colors. Additionally, a deep blue band with a symmetrical pattern appears just beyond the outer boundary of the healthy living being. A fingertip presents an image with these characteristics.
- On other hand, if we photograph the fingertip of a person who has reported feeling tense, aroused, or anxious, the registered pattern appears as a luminous red cloud or blotch.

- Paranormal researchers say that the qualities of the aura can be linked to the health of the living being who is being examined and can be used to diagnose diseases. These qualities are also said to be useful for determining mental states or paranormal capabilities of a person.
- Another important aspect of Kirlian photography is artistic. The charges leaving the leaves and other objects draw beautiful figures. Exhibitions of Kirlian photos taken of various objects are common.

In the next pages, we will describe some projects based on high-voltage devices that can be used as Kirlian cameras and in other paranormal experiments. These circuits are based on the first Kirlian machines built by the author many years ago and updated to use modern components. Some of these projects are reprinted from the Brazilian magazines *(Revista Saber Eletrônica* and *Eletrônica Total)* with which the author is employed.

2.7.3 Experimenting with Kirlian Photography

If the reader wants to start with experiments involving Kirlian photography, some basic equipment is necessary, and details about the basic procedures are important, as they are in other paranormal fields. As in the case of EIP and EVP experiments, some caution is required to keep the researcher from tricking himself. Because of our tendency to extrapolate results, false conclusions can also occur when experimenting with the Kirlian effect.

Before assembling the circuits and starting with the experiments, familiarize yourself with scientific explanations for some key concepts such as corona effect, ionization, and high-voltage fields. Take care when choosing your sources of information about the subjects, verifying their origin and making sure that the author knows enough about the subject and is not merely trying to sell you his ideas.

Basic Materials

1. *Kirlian machine.* This is the basic equipment for all related experiments. The Kirlian machine is a high-voltage generator with special characteristics, producing a voltage between 8,000 and 50,000 V, with a very low current output and operating frequencies between 100 and 10,000 Hz. The main point to be considered about this machine is whether the design is safe. Working with very high voltage is dangerous, and a shock can be fatal under certain conditions. The reader who wants to experiment with a Kirlian machine must become familiar with its operation before starting the experiments. Some Kirlian machine circuits will be described in the following pages.
2. *Photographic paper or other media for registering the images.* There are many ways to record the images produced by objects in Kirlian machines. Images of fingertips, plant leaves, coins, and insects can be registered on photosensitive materials, fax paper, and any media sensitive to light or electricity.

The reader is free to experiment with other media. Polaroid cameras and film have been used successfully to make Kirlian photos.
3. *Magnifying glass.* Details of the "aura" can be observed more easily with the aid of a magnifying glass.
4. *Scanner and computer.* The advanced experimenter can use the computer to edit the images. You can scan the images and process them using standard image editing software.
5. *Materials for processing film and prints.* This includes chemistry for developing photographic materials, related equipment, a darkroom, etc.

A Note about the Power of the Practical Projects. Again, we emphasize that *working with high voltages is dangerous.* Although this book provides all the information the reader needs to conduct safe experiments with Kirlian photography, as the output power of the Kirlian machines increases, the danger of mortal shock rises in the same proportion. Therefore, to ensure the readers' safety, the projects described herein, although possibly dangerous in some conditions, are limited in power. The reader can alter some of them to increase the output power and thereby obtain auras of larger objects such as a complete hand. However, we don't recommend that any reader do so immediately. First, build a low-power unit to become familiar with the procedures for using it. Only after you are certain that you have the required expertise will it be safe to work with more powerful units.

Project 22: High-Voltage Generator (Kirlian Machine I)

Our first project can produce high voltages between 5,000 and 30,000 V using a common car ignition coil. This circuit produces a low-frequency, pulsed, high voltage that can be used to see or photograph auras in small objects such as coins, leaves, insects, biological samples, and even the fingers of a human. Many researcher say that very low frequencies are not the best for seeing human auras and use medium or high frequencies in the range of 2,000 and 10,000 Hz.

In fact, very low-frequency signals produce short-duration sparks that can cause some sensation of being shocked. If this occurs, replace the glass or plastic in the electrode by a thicker one. It is normal to feel a slight heat or tingling sensation when placing your fingers on the electrodes.

The circuit is powered from the ac power line, and care must be taken when using it to avoid shock hazards. The use of an isolation transformer between the circuit and the ac power line is important.

Important: Do not touch any part of the circuit when it is in operation, as it is plugged directly to ac power line! Always use an isolation transformer.

When photographing your fingers or hands, always wear shoes. Allowing your feet to contact the ground directly can increase the currents through your body, causing severe shocks. If you feel a sensation of being shocked, replace the glass sheet with a thicker one.

How It Works

The circuit consists of a relaxation oscillator using a neon lamp and a silicon-controlled rectifier (SCR). The ac power line voltage charges C1 and, at the same time, C2 via R1 and P1. As soon as the voltage across C2 reaches the ignition point of the neon lamp, a conduction pulse is produced and applied to the gate of the SCR.

With the presence of the pulse, the SCR triggers on, closing the circuit formed by C1 and the low-voltage winding of the ignition coil. This means that C1 discharges via the ignition coil, producing a high-current pulse. In consequence, a high-voltage pulse is produced in the secondary winding of the ignition coil.

With the discharge of C1 and C2, the neon lamp turns off, and with it the SCR. A new charging cycle begins, and the described process repeats.

The charge and discharge cycle can be adjusted by P1. With the values shown in the circuit, the oscillator will operate at a frequency between 100 and 1,000 Hz. Reducing C1 and C2 to 1 µF and 0.047 µF, the frequency can be increased, but the output power is lowered.

The high voltage induced in the ignition coil depends on the voltage reached by the C1 charge in each cycle. This voltage can rise to values between 4,000 and more than 30,000 V.

As the current in the secondary winding of the ignition coil is very low, the high voltage is not life threatening, but it can cause a severe shock to anyone who touches the line. Note that there is no isolation between the power supply line and the output of the high-voltage circuit. This means that special care must be taken when using the device. For more secure operation, an isolation transformer on the input is recommended.

Assembly

The schematic diagram of the high-voltage circuit for Kirlian photography in its first version is shown in Fig. 105. R1 is a wire-wound resistor. This resistor can be altered according to the ac power line voltage. The values in parentheses are for the 220/240 Vac power line. Electrolytic capacitor C1 must have values between 2 and 16 µF. This capacitor determines the amount of energy delivered to the ignition coil in each pulse and therefore the output power of the circuit.

Any ignition coil such as the ones commonly used in cars can be used. All the types will function, and the only difference is the voltage found in the output, which can vary within a wide band of values.

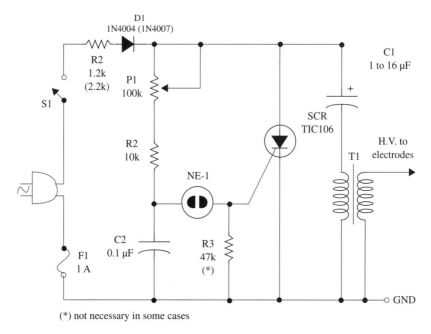

Figure 105 Kirlian machine I.

The SCR must be mounted on a small heatsink. A piece of metal can be used as a heatsink. Types with suffix D or E are recommended if you are plugging the circuit into a 117 Vac power line. For a 220 or 240 Vac line, the recommended SCR will have the suffix E.

C1 is any electrolytic or polyester capacitor rated to 250 WVDC or more. If the circuit is connected to a 220 Vac or 240 Vac power line, the capacitor must be rated to 400 WVDC or more.

The circuit must be installed in a wooden or plastic box to avoid contact with living things, which can cause a severe shock hazard. **Do not use a metallic box.**

T1 is any isolation transformer with a turn ratio of 1:1. The primary winding is rated to the ac power line voltage. Output power is rated to at least 20 W.

Warning: Make tests with your multimeter to ensure that the isolation between windings is high enough (2 MΩ or more) to guarantee safe operation of the circuit.

Important: See earlier precautions related to safe use of the machine.

The Electrode

A metal plate (copper, aluminum, or other metal foil) with dimensions from 10 × 10 cm to 15 × 15 cm is placed onto a plastic base or glass base several centimeters larger, as shown in Fig. 106. You can use any glue to affix the electrode, or even adhesive tape.

You can use common wire to connect the electrode to the high-voltage transformer if it is not placed at too great a distance. If the electrode is placed far from the circuit, the wire that connects the plate to the circuit must be of a special type as used to connect the high-voltage circuits of TVs. This is applicable to distances above 50 cm. Common plastic-covered wires do not completely isolate the high voltage produced by the circuit and can allow dangerous shocks if touched.

Over the electrode, place a piece of glass or plastic. The thickness of this glass piece depends on the high voltage generated by the circuit. Pieces with a thickness between 4 and 10 mm are recommended. Tests must be made to determine which thickness affords the best results without allowing the sensation of being shocked.

This electrode can be used with all other high-voltage circuits described from here on. Of course, experiments must be repeated in each case to determine the correct thickness of the glass or plastic sheet.

Using a TV High-Voltage Transformer

The ignition coil can be replaced by a horizontal output transformer as found in TV receivers or computer monitors. This transformer has some advantages over the common ignition coil in this project.

If you have an old nonfunctional TV (black-and-white or color), you can remove from it a horizontal output transformer that can be used for the high-voltage electrodes. The high-voltage transformer is used in TVs and computer monitors to produce voltages between 5,000 and 35,000 V, which are needed to accelerate the electron beam for producing images in the screen of the cathode ray tube (CRT). You only have to take off the transformer and add 20 to 40 turns of common wire (AWG 22) to the core as shown in Fig. 107.

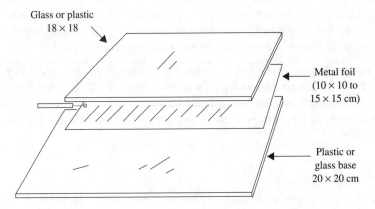

Figure 106 The electrode.

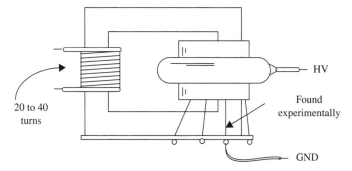

Figure 107 Using a high-voltage TV transformer.

The high voltage to the electrodes is taken from the upper terminal of the coil. The other terminals correspond to taps in the coil, and you must find the one that gives the best results when attached to the ground.

An important advantage is gained by the use of this kind of transformer as a replacement for the ignition coil: the high-voltage winding is completely isolated from the circuit and so from the ac power line. You will not need the isolation transformer in this case.

If, when using this transformer, sparks are produced in the windings and no high voltage is detected at the output, the transformer probably was affected by humidity or has isolation problems. You can try to leave it in a dry place over a period of time to eliminate the humidity and then try again. If this does not work, the transformer can't be used in this project.

Although this doesn't eliminate the possibility of getting a severe shock if you touch a high-voltage part of the electrode, the danger of receiving a direct discharge from the ac power line is eliminated.

High-Voltage Detector

A neon lamp can be used to detect the high voltage produced by this circuit and others described in this book. This lamp is mounted at the end of a plastic tube (a spherical pen tube) as shown in Fig. 108. If you place the neon lamp near the

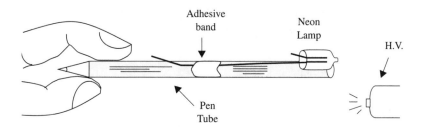

Figure 108 The high-voltage detector.

electrode or output of the high-voltage transformer, it will glow with an orange light. This neon lamp test can be used in all the experiments with Kirlian machines as an efficient high-voltage detector.

Any fluorescent lamp can also be used to detect high voltage. Touch one of its terminals to the electrode or the high-voltage output of the transformer. The lamp will glow. With very powerful high-voltage sources, the lamp glows if it is simply near the source.

Connect the ground terminal to any suitable ground source. Ground sources include the ground terminal of the ac power line outlet or any large metal object that makes electrical contact with the Earth.

Warning: If the ground terminal is not connected to a proper grounding point, the output power will be reduced, and the results when observing or photographing auras will be inferior. However, when the ground connection is made, the circuit becomes more dangerous as the output power is increased.

Using the Circuit

Initial experiments can be made using common objects and living things such as leaves, coins, paper clips, insects, keys, etc., as samples. Turning on the circuit, you'll hear the sound of the high-voltage ions escaping from the electrode. This sound is like a "hiss." A smell of ozone can be noted during the operation of the high-voltage sources (ozone is produced when three biatomic molecules of atmospheric oxygen are converted into two triatomic oxygen molecules, or ozone). The aura can be seen in the dark when the samples are placed onto the electrodes.

Adjust P1 to get the best aura. This potentiometer controls the high voltage applied to the electrodes. Don't touch the electrodes during these experiments, even though you have determined that your project is safe.

To see the aura of your fingers, adjust P1 to get the lowest power (use a coin as sample) and then put your finger onto the electrode. The normal reaction is to feel a little bit of heat. If you feel a shock, you must replace the glass piece with a thicker one or adjust P1 to reduce the voltage in the electrodes. You can also make changes in the frequency, increasing it to reduce the shock sensation. Adjust P1 to get the best image without shock.

Important: Take care when making these experiments, as the circuit is wired to the ac power line.

After some experimentation with visual images, you can try to make photos or register the images on paper.

Observation

How the pattern you saw in the dark is registered on fax paper or in a photograph depends on some characteristics of the Kirlian machine. This circuit produces low-frequency pulses that are responsible for sparks or bright ion streams leaving the sample. This means that the image is different from those produced by higher-frequency machines such as described in the following projects. Of course, you can alter the operational frequency as explained below.

Using Fax Paper. Cut a small piece of fax paper (5 × 5 cm or according to the size of the object to be photographed) and place it on the electrode, and place an object on the piece of paper (a coin, for instance) as shown in Fig. 109. Make sure that the sensitive side of the paper is in contact with the object. If the paper is inverted, the process doesn't work.

Turn on S1 and adjust P1 to get the best performance. You'll see the aura of the object and hear the "hiss" of the charges leaving the object. Wait between 5 and 15 seconds. Turn off the device and reveal the photo. The best time interval will be found experimentally according to the fax paper and the power of the Kirlian machine. Depending on the paper sensitivity and the output power of the Kirlian machine the image may appear on the paper after an interval of time without the need of any kind of processing. But the image can be enhanced as shown in Fig. 110. Iron the paper, adjusting the iron to a very low temperature. If the iron is too hot, the image is burned and will appear as a black spot. The time and the temperature must be also found experimentally according to the fax paper, the type of

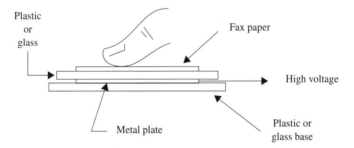

Figure 109 Using fax paper to register the aura of a fingertip.

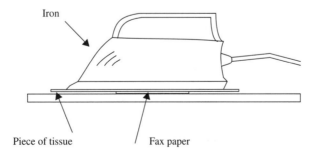

Figure 110 Revealing the fax paper image.

iron, and the temperature. A fax image generated by the author appears in Fig. 111.

Using Photographic Film. Working with photographic film requires some extra care. There are several techniques in use by researchers.

Working in the Dark. Place a piece of film between the plate and the object to be photographed. If you are photographing your fingers, Fig. 109 shows how the film is placed. Simply place the film in the location shown for the fax paper.

Turn on the high-voltage circuit (Kirlian machine) for 3 to 20 seconds. The best time interval will be determined by experimentation and can vary according the sensitivity of the film and the output power of the circuit.

When photographing metallic objects such as a coin, or living protoplasm such as a leaf, it is important to install a ground electrode as shown in Fig. 112 to increase the current and the corona effect.

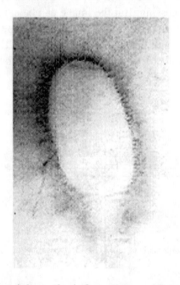

Figure 111 Aura image of the author's fingertip on a piece of fax paper, scanned for placement in this book. The Kirlian machine used to make this image was the one described in Project 22.

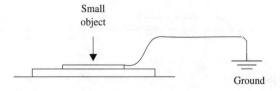

Figure 112 Grounding small objects.

After exposure, the film must be placed in an opaque, light-proof envelope before you can turn the lights on again. The film is ready for development.

As the output power of each circuit can vary within a large range as determined by the components' tolerances, and the film characteristics can affect the image, the experimenter must make several photos to find the correct exposure time and adjustment of P1.

Working in the Light. It is possible to make photos under ambient illumination if you protect the film with opaque cards. In a dark ambient, place the piece of film between the two opaque cards or in a light-proof envelope. Then, in the ambient light, you place the envelope containing the film between the object and the electrodes as shown by Fig. 113.

Turn on the high-voltage generator and expose the film from 3 to 30 seconds to register the image. Turn off the device and process the film. Depending on the film sensitivity (ASA or DIN), the correct exposure time will vary and must be determined experimentally.

Using Polaroid Film. Polaroid film provides another good option for producing Kirlian photos or small objects.

- *Working in the Dark.* The room where the experiment is made must be totally dark. Place the film on the electrode, and on the film place the object to be photographed (a coin, your finger, etc.).

  Turn on the circuit (which should already have been adjusted to give the best results with the object to be photographed when viewing the images) for a time duration of 1 to 10 seconds. Again, the correct interval must be found experimentally, as it depends on the circuit characteristics (which change with the tolerances of the components) and the sensitivity of the film. Polaroid films such as the 600, SX-70, or Time Zero are good. We observe that Polaroid 600 is not a high-definition film and will result in blurred images.

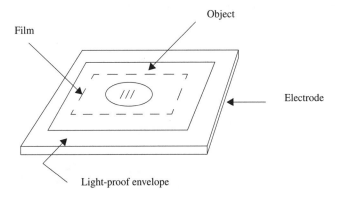

Figure 113 Working with photographic films.

After exposing the film, place it back in the camera (in the dark) and take a photo pointing the camera at any dark place. (Do not use a flash.) Then, wait for the print to be ejected from the camera.

Caution: Polaroid films have a metallic white border that is conductive, causing arcs and electric shocks if touched. When using these films, cut a window in a piece of plastic to make a mask as shown in Figure 114.

- *Working under Normal Light.* You can use your Polaroid camera to expose the film by putting it inside a light-proof wood or cardboard box. This box must be large enough to encase the Polaroid camera and have some additional space for working with the film. Using a black or very dark fabric, make two sleeves with elastic on one end, similar to coat sleeves. The sleeves are attached to two holes made in the box. Use staples, glue, or other suitable means to fix the sleeves. It is important to make the sleeves and holes light tight. You can handle the film inside the box, putting it on the electrode and then placing the object to be photographed on top of it. If you are trying to photograph your fingertips, put it on the electrode. Turn on the Kirlian machine for 1 to 10 seconds, according to the film speed. Then, place the film in the camera and take a photo in the dark as before so that the print is ejected.

You can also use this dark box to insert film into a light-proof envelope. You then take it out of the box, take the photo, and place it back into the box for later processing.

Suggestions

- Use color filters between the film and the object to be photographed. Try filters created by your imagination such as color cellophane pieces and filters such as the ones described for EIP experiments (Part 1).
- Photo films that are sensitive to UV (ultraviolet) or IR (infrared) can be used for experimentation. It is worth noting that a wide assortment of filters and op-

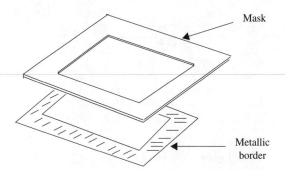

Figure 114 Working with Polaroid films.

tical materials can be purchased from Edmund Scientific Corporation (web site http://www.edsci.com).
- Test other kinds of light-sensitive materials such as photostatic paper. The images of auras can also be seen with the aid of night-vision binoculars.
- Scan the images and process them using a computer. Many programs can be used to process the images, revealing details not seen in the original images.
- The frequency can be increased by reducing R1 and C1. (The lowest recommended value for R1 is 470 Ω in the 117 Vac power line and 1 kΩ in the 220/240 Vac power line.) The value range for C1 is given in the parts list.

Photographing Small Objects. If small objects are placed on the electrode of a Kirlian machine, they can't produce a good photo unless they are connected to ground. To connect the objects to the ground, you may use a wire with alligator clips on the ends. In the case of a leaf, choose a specimen that is no bigger than half the size of the electrodes (2.5 to 5 cm long). Connect the ground wire directly to the tip of the stem or leaf. Use an alligator clip for this task (try to find an alligator clip that does not press too hard on the leaf, cutting its tip). Cover the leaf with a small piece of glass or plastic to hold it flat on the film. Experiments will show you how long you have to power up the circuit and how much pressure must be exerted on the leaf to get a good image.

If you are photographing a coin, the plastic or glass cover is not required, since the coin is heavy and flat enough to lie as desired on the electrode.

Interferences. Many of the circuits described herein can radiate spurious signals in the radio spectrum used for communication services such as TV, radio, etc., interfering with many communication devices. To avoid interference, do not use the Kirlian machines near such equipment.

The Phantom Leaf. One of the most important phenomena found by Kirlian photography is referred to as the "Phantom Leaf Experiment." Cut about 2 cm off the tip of a leaf and photograph it. If you are lucky, you'll get an image of a whole leaf, even though the tip is missing!

Parts List: Project 22

Semiconductors

D1	1N4004 or equivalent silicon rectifier diode (use the 1N4007 in 220 or 240 Vac power lines)
SCR	TIC106-D silicon-controlled rectifier

Resistors

R1	1,200 Ω or 1,500 Ω, 10W, wire-wound resistor (2,200 Ω × 10 W if the circuit is used in a 220/240 Vac power line)
R2	10 kΩ, 1/8W, 5%—brown, black, orange

Capacitors

C1   1 to 16 µF/250 WVDC, electrolytic or metal film (see text for the 220 or 240 ac power line)

C2   0.1 µF/100 WVDC or more, ceramic or metal film

Miscellaneous

F1   1 A fuse

NE–1   NE–2H, any common neon lamp

P1   100 kΩ potentiometer

S1   SPST, toggle or slide switch

T1   Ignition coil or high-voltage TV transformer (horizontal transformer) (see text)

Printed circuit board, plastic or wood box, knob for P1, wires, electrodes, solder, etc.

Project 23: High-Voltage Generator (Kirlian Machine II)

Our second circuit for a *Kirlian camera* or *Kirlian machine* uses a common automotive neon light transformer. These devices are not really transformers but inverters (or dc/ac converters) that change the dc voltage of 12–14 V of a car battery to high-voltage ac of 2,000–4,000 V. The device is basically formed by a high-power oscillator driving a high voltage transformer.

In particular, we recommend for this project three high-voltage neon transformers that can be found in the U.S. market and that furnish output voltages between 2,000 and 4,000 V.

One is MINIMAX4, from Amazing Products, a division of Information Unlimited (http://www.amazing1.com/voltage.html). It is rated to 14 Vdc of input voltage and an output of 3,000 to 4,000 V, with current between 5 and 10 mA. Another device from the same company, but less powerful, is the MINIMAX3, with specified output voltages between 2,000 and 3,000 V.

The third is the NEONXA, also from Amazing Products, and information can be found on the internet page http://amazing1.com/neondc.htm.

The main advantage of using these transformers is that the circuit is easy to mount. The reader will only have to add the power supply (or use a car battery or even 8 D cells) and the electrode.

In our project, we are going to power this circuit from a 12 Vdc power supply and plug it into the electrodes described in the previous projects.

As the frequency used for this type of inverter is higher, and the performance is also better than that of the previous project, good photos and images of auras can be obtained.

How It Works

The circuit is formed by a 12 V power supply, the car neon lamp transformer, and the electrode. For the power supply, a simple circuit is given. The ac voltage is lowered by the transformer and applied to a diode bridge. The low dc voltage after the diodes is filtered by C1 and stabilized at 12 V by the 7812 integrated circuit (voltage regulator).

The 12 Vdc is used to power the car neon lamp transformer that furnishes an output between 3,000 and 4,000 V. This high voltage is applied to the electrodes.

An option for the power supply is the use of a car or motorcycle 12 V battery. It is also possible to use 8 D cells, or two 6 V lantern batteries wired in series. But because the current drain is high, the life of the batteries is not long when used for this task.

Another option is the use of a 12–14 V power supply plugged to the ac power line (commercial type). Find one rated to 800 mA or more, and observe the output connector polarity (positive or negative in the center as shown in Fig. 115).

Assembly

Figure 116 shows the complete diagram of this version of a Kirlian machine using a car neon light transformer. The power supply is the only part to be mounted by the reader. A terminal strip can be used as the chassis for the small components of this part as shown in Fig. 117.

It is important to install all the parts in a box, as we are working with ac power line voltages, which present a real danger of shock if any live part is touched. We recommend the use of a plastic or wooden box to reduce the danger of shock if any internal part touches it. *Do not use a metallic box.*

Electrodes

The electrodes are mounted as in the previous project. Use short wires to connect the output of the high voltage transformer to them. The electrodes can be placed on the box.

Testing and Using the Circuit

The neon lamp high-voltage detector can be used to test this circuit. If the circuit is OK, placing the high-voltage detector near the electrode will light the neon lamp.

Figure 115 Outputs of ac/dc converters.

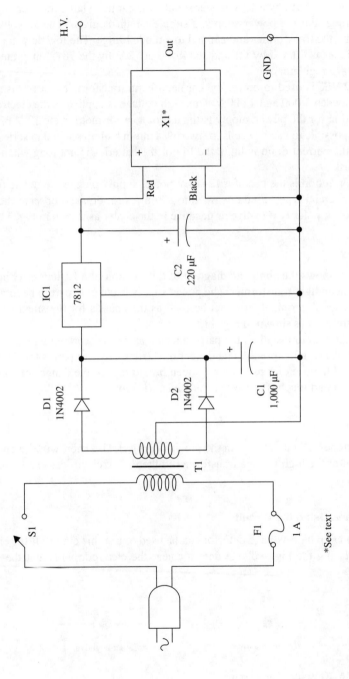

Figure 116 Kirlian machine II.

Experimenting with Images 149

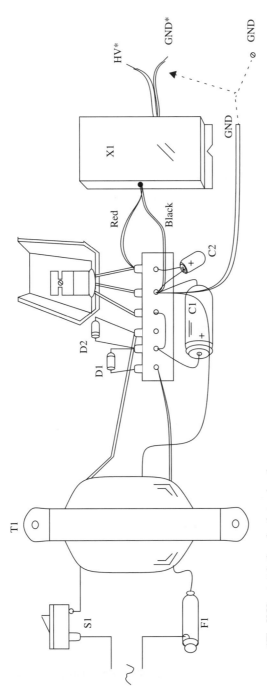

*The H.V. output is the wire that makes the longest spark to the ground.

Figure 117 A terminal strip used as the chassis.

Connect the ground terminal to a suitable ground source. Turn on S1, powering up the circuit. A "hiss" can be heard if the circuit is operating properly. This is produced by the high voltage in the electrodes.

In the dark, put your finger on the electrode. If you feel a sensation of being shocked, you need to make some changes in your project, as follows:

1. Replace the plastic or glass on the electrode with a thicker piece.
2. Make sure that the transformer is the correct one and that it is wired as indicated in the project description.

If the circuit seems to work well, you can start with your experiments.

Always wear shoes when making these experiments, and do not touch any metallic object. If plants or other objects with no connection to ground are used in the experiment, connect them to ground with a wire.

Suggestions

- Use color filters as in the other experiments.
- Films sensitive to UV and IR (ultraviolet and infrared) can be used experimentally for registering the images.
- Capture the images registered in the photos using a scanner and process them using any suitable computer software.

Parts List: Project 23

Semiconductors

IC1	7812 integrated circuit, voltage regulator
D1, D2	1N4002 or equivalent silicon rectifier diodes

Capacitors

C1	1,000 µF/25 WVDC, electrolytic
C2	220 µF/16 WVDC, electrolytic

Miscellaneous

T1	Transformer, 117 Vac or ac power line voltage primary; 12 C, CT × 1 A secondary
F1	1A fuse
X1	Automotive neon light transformer, MINMAX4 or NEONXA (see text)

Terminal strip, heatsink for IC1, wires, fuse holder, electrodes, photographic films, power cord, solder, etc.

Project 24: High-Voltage, High-Power Generator (Kirlian Machine III)

This is the most powerful generator of the three circuits described in this book. Voltages to 40,000 V, and even higher, can be produced and applied to the electrodes described in the previous projects. *You may need to use thicker glass or plastic pieces to avoid being shocked by the higher voltages produced by this circuit.* Although this is a high-power circuit, the current is very low, which reduces the danger of mortal shocks if any part of the circuit is touched. Of course, *extreme care* must be taken when working with high-voltage circuits, as there is always a risk of shock.

Warning!
High voltages are dangerous. Take extreme care when working with this device. Previous experience with this kind of device is required.

The reader can use this circuit in experiments involving Kirlian photography (electrophotography) and in other paranormal experiments such as those involving plasma or ionization. The output power of the circuit is up to 20 W, and common components are used.

Some characteristics of the circuit are as follows:

- Power supply voltage: 117 or 220/240 Vac (ac power line)
- Output voltage: up to 40 kV (depending on the high-voltage transformer)
- Output power: 5 to 25 W (depending on the components used)
- Number of transistors: 1
- Operating frequency: 2 to 15 kHz

How It Works

The circuit is formed by a one-transistor Hartley oscillator with the operating frequency determined by C3, C4, and the inductance of the primary side of the high-voltage transformer.

A high-power NPN silicon transistor is used in the project. This transistor must be mounted on a large heatsink due the high current drained from the supply.

R1 and R2 fix the output power by controlling the current drain of the transistor. R3 is the base bias. Depending on the characteristics of the transistor, which will vary even between units of the same type, resistor R3 must be experimentally specified in the range between 270 and 470 Ω.

The high-voltage transformer, which also determines the operating frequency, is a TV horizontal output transformer with ferrite core. The primary is formed by 20 to 40 turns of common plastic-covered wire. In the secondary of this transformer, there appears a very high voltage that will be used in the experiments.

The power supply is simple, as regulated voltage is not needed. A 25 + 25 V transformer is recommended, but the voltage is not critical. Types with secondary windings rated to voltages between 20 and 25 V and currents between 3 and 5 A can be used experimentally.

Assembly

The schematic diagram of the Kirlian machine III is shown in Fig. 118. As the circuit is not critical, an alternative mounting technique is suggested in Fig. 119. A terminal strip is used as the chassis. This terminal strip holds the small components, such as the resistors and capacitors, which are connected to the other components by wires. Heavy components such as the transformer are affixed by screws directly to the box.

The terminal strip and the other components can be housed in a plastic or wooden box. As recommended in the other high-voltage projects, to avoid problems of shorts or shocks, *do not use a metal box.*

The high-voltage transformer can be cannibalized from any old non-functioning black-and-white or color TV. If possible, find a 21 inch or larger TV, as the high-voltage transformer used in these units can supply higher voltages.

R1 and R2 are wired-wound resistors. C1 is not critical, and values between 1,500 and 4,700 µF can be used.

The primary winding is made with common plastic-covered 22 or 24 wire (rigid or flexible) as shown in the figures, with 35 + 35 turns wound in the core. Make sure the wire is wound in the same direction after the tap.

The transistor must be mounted on a large heatsink. Keep all the wires short to avoid oscillations and other problems.

Experimenting

Plug the power cord into the ac power line. Turn on S1, keeping S2 open. The circuit will produce a light tone or hiss indicating oscillation.

Warning:
Do not touch any part of the circuit when testing!

Use a neon lamp as in the previous projects. Placing the neon lamp near the high-voltage transformer causes the lamp to glow. This indicates that high voltage is being produced.

You can also use a common fluorescent lamp to detect the high voltage. Take a fluorescent lamp (even one that doesn't work anymore when plugged into the ac power line) and place it near the high-voltage terminal of the high-voltage transformer.

If the transistor overheats, it is a good idea to replace the heatsink with a larger unit. Resistors R1 and R2 operate hot; don't be concerned about that.

The electrodes are constructed as in Project 20. The glass must be thicker in this project due to the very high voltage. Perform experiments with glass or plastic electrodes at least 3 mm thick.

Experimenting with Images   153

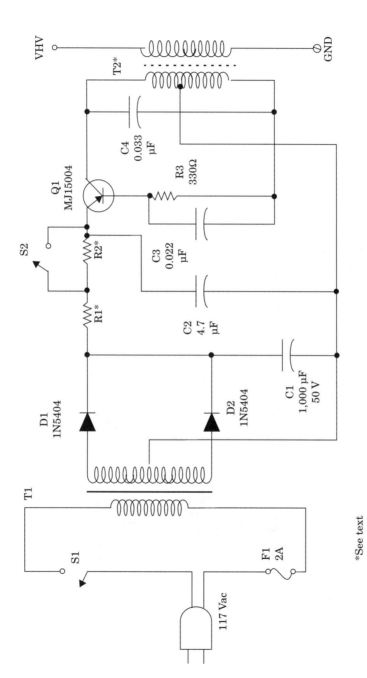

*See text

Figure 118   Kirlian machine III

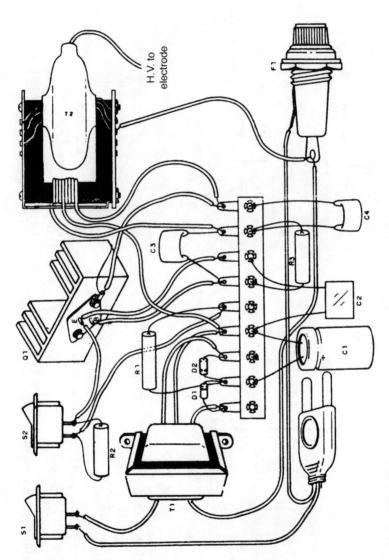

Figure 119 Assembly using a terminal strip as chassis.

An interesting experiment is made by placing a high-power incandescent lamp at the high-voltage output as shown in Fig. 120. Common incandescent lamps are filled with an inert gas to eliminate oxygen, which would burn the filament. This is required to avoid a vacuum in the lamp, which would create an implosion in case of any impact.

Exposed to a high-voltage field, the inert gas ionizes, and you will see beams of ionized gas emanating from the filament as in a plasma lamp. The filament will not heat in this circuit, as no current flows through it. The current flows from it to the glass and from there to the air as electric charges. This is the effect you see in the common plasma lamps.

In our project, we can use any common incandescent, white, glass lamps for the ac power line (117 or 220/240 Vac) with power ratings from 150 to 500 W.

Put your finger on the glass (take care!), and you'll see a flux of ions and a luminescence involving it (aura).

S2 will increase the output power of the circuit.

If the circuit doesn't operate as expected, due the characteristic differences between the recommended and installed components (due to tolerance variations and other factors and, in particular, if Q1 heats too much), try the following modifications:

- Vary R3 in the range between 270 and 470 Ω.
- Vary R1 and R2, increasing them to 4.7 or even 5.6 Ω.
- Vary C4 in the range between 0.027 and 0.1 µF
- Vary C3 in the range between 0.015 and 0.1 µF
- Vary the number of turns of the primary winding of T2 from 20 + 20 to 50 + 50.

Suggestions

- Use this circuit to power a fluorescent lamp with a high-frequency/voltage signal, and make experiments picking up images or sounds (with the light-to-sound converter).

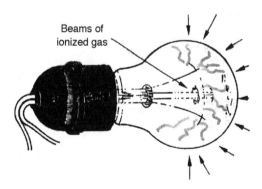

Figure 120 Plasma lamp using a common incandescent bulb.

- A wire-wound potentiometer (470 Ω or 1 k Ω × 5 W) can be wired in series with R3 to control the output power, and also the frequency, of the high voltage applied to the electrodes.
- Figure 121 shows how a high voltage rectifier and a glass capacitor can be made to produce a high dc voltage to experiments.
- If a nail is placed in the output voltage terminal with the glass capacitor and the high-voltage rectifier, as shown by Fig. 122, the circuit can be used to produce negative (or positive) ions in the air.

Parts List: Project 24

Semiconductors

Q1   MJ15004 high-power NPN silicon transistor

D1, D2   1N5404 silicon rectifier diodes

Resistors

R1, R2   3, 3 Ω × 10 W, wire-wound

R3   330 Ω × 5 W

Capacitors

C1   2,200 µF/50 V, electrolytic

C2   4.7 µF/200 V, polyester or ceramic

C3   0.022 µF/200 V, polyester or ceramic

C4   0.033 µF/200 V, polyester or ceramic

Miscellaneous

F1   2A fuse

S1, S2   SPST, toggle or slide switches

T1   Transformer: primary 117 Vac or 220/240 Vac, secondary 30 + 30 V × 2 A (see text)

T2   High-voltage transformer (see text)

Plastic or wooden box, fuse holder, electrodes, terminal strip, wires, solder, power cord, etc.

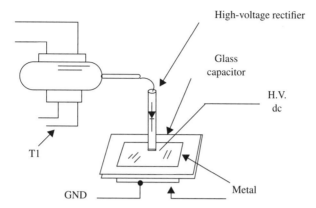

Figure 121 Producing dc high voltage.

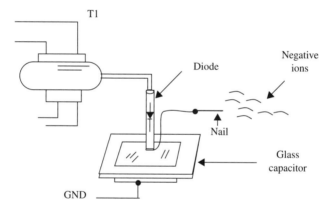

Figure 122 Producing negative ions.

Part 3
Experimenting with Paranormal Skills

A mind not to be changed by place or time.
The mind is its own place, and in itself
Can make a Heaven of Hell, a Hell of Heaven.

John Milton (1608–1674)

Paranormal skills are not the privilege of a restricted group of individuals. Everyone, including you, has latent paranormal abilities. But discovering and developing your paranormal skills is not easy.

Many techniques can be used to find and develop such abilities as seeing into the future or moving objects using the mind. Most of them use some kind of ritual that has been handed down through generations and generations of people who do not have any formal knowledge about science.

Even living in today's world of high technology, the paranormal skills are studied and analyzed from a traditional, mystic angle. Mysticism and religion are more commonly involved than true science, but science can explain many concepts more easily and without the "aura" of mystery. The greatest problem for the serious paranormal phenomena researcher, as we mentioned in other parts of this book, is to separate what is real from what is "noise" and apply the correct concepts of modern science to explain what is really happening.

The natural human tendency to use technical terms such as *energy, fields, vibrations,* and others in our attempts to add a scientific aspect to many paranormal phenomena is the main obstacle to making "official" science accept the theories and hypotheses created by researchers.

As we have mentioned, this book is not based on an acceptance of all paranormal phenomena, which are diverse, nor is it intended to explain any of these phenomena. Our purpose is to discuss topics that the researcher may believe to be worthy of further research and to show how electronics can provide some practical aids in exploring these subjects. This implies the possibility of adding some new techniques to a field of experimentation in which most of the technology dates back to Medieval times or earlier. Successful research in the field of paranormal phenomena requires not only skilled individuals but some technical assistance as well.

The following pages are dedicated to introducing the reader to some devices that can be used in experiments involving individual paranormal skills such as extrasensory perception (ESP), telekinesis or psychokinesis, clairvoyance, reaver, far-touch, transcendental meditation, radiesthesia, and others.

3.1 Paranormal Skills

Many researchers in the field of paranormal phenomena believe that our mental processes are not limited to the volume of the brain. All the processes running inside our brains, and not only when we are in a conscious state, involve not only interaction with other parts of our body but also some kind of unknown interaction with the surrounding space, perhaps extending throughout the world and all the universe. Our brain operates like a "radio transmitter/receiver" *(transceiver)*, sending and receiving information directly from any part of the universe. (Again, we caution the reader not to use these technical terms with the meaning they have in "official science." The quotation marks indicate the figurative use of the term.) This means that our common senses of hearing, sight, smell, taste, and touch are not the only ways we have to interact with the world and the universe.

Many unknown processes interact with our normal senses, and sometimes they appear as large-scale manifestations in some individuals. These manifestations, being inexplicable by "official science," are called *paranormal,* and the individuals are placed in a category of humans referred to as *sensitives*.

Everybody is sensitive, but to different degrees. What is your degree of sensitivity, and to what kind of phenomena? This can be a very interesting field of research for you, and it can be determined by simple experiments.

Can you move objects using your mind? Can you see distant places without using your eyes? Can you preview what number will be chosen in the next lottery? Can you control your physiological functions using your mind? Can you know what somebody else is thinking?

Extrasensory perception, telekinesis, psychometry, clairvoyance, radiesthesia, psychokinesis, and far-touch are examples of the paranormal skills revealed in individuals who, in some manner, can "tune" some information from the beyond, or from some unknown place in an unknown dimension, and use it as a "sixth sense."

If you don't believe that you have any chance of being a paranormal individual, it is probably because you have never stopped to think about some revealing facts that occurred in your life.

- Have you ever, when walking alone late at night in a deserted street, felt the clear sensation of the presence of an invisible someone (or something) walking beside or behind you?
- Have you ever, when visiting a distant place where you had never been before, had the strange sensation that the place was familiar and you had visited there before (referred to as *déjà vu* in French)?
- Have you ever felt the sensation of seeing a member of your family appearing in a doorway for a second, even though that person is a thousand miles away, and soon afterward received notice that the person had died?
- Have you experienced strange phenomena in your house, such as items falling (e.g., the sound of rocks falling on the ceiling) without any explanation, and later received notice of some accident involving a relative?

- Have you noticed your dog searching or barking at a place in your house where there seems to be nothing, indicating that the animal can see somebody or something that is invisible to you?
- Have you seen strange lights or flying objects while traveling?

These common occurrences constitute evidence that not only *sensitives* but everyone sometimes can tune into the paranormal and extend their sensory perception into who-knows-where, or even who-knows-when, to extract information from the "files of the beyond." There are many more paranormal skills to be studied in humans than we have yet imagined.

Before starting with the experiments using electronic devices, it is important to provide the reader with some important information about the terms used for paranormal phenomena and their meanings.

3.2 Paranormal Phenomena

3.2.1 Extrasensory Perception or Clairvoyance

ESP is defined as the ability to sense (feel) or see (in the *mind's eye*) things (places and people) that are far away. It is also called the *sixth sense*. It refers to an individual's ability to receive information from beyond the ordinary five senses of sight, smell, hearing, taste, and touch. This individual can be provided with information not only of the present but also from the past and the future. It seems that the information comes from a *second or alternate reality*.

History. ESP manifestations have been related since biblical times. The first person to use the term *extrasensory perception (ESP)* was Sir Richard Burton, in 1870. In 1892, another researcher, Dr. Paul Joire, used the term to describe the abilities of a person who had been hypnotized or was otherwise in a trance state to externally sense things without using the common senses.

The first systematic research on ESP was conducted by the Society for Psychical Research, in London. Similar research was soon carried out in other countries, including the U.S.A. However, these first studies were rarely experimental. They mostly examined spontaneous incidents in field (uncontrolled) conditions. Only rarely were they examined under laboratory condition as we do today.

In 1930, at Duke University (U.S.A.), Dr. J. B. Rhine began conducting studies of psychic phenomena in the university's psychology department. Rhine was the first experimenter to perform psychic testing using Zener cards (cards with the five symbols of circle, square, star, plus sign, and wavy lines, as shown in Fig. 123), which were developed for him by his colleague, Dr. Karl Zener. The symbols were printed singly, in black ink, on cards resembling playing cards.

Rhine took his studies away from Duke University in 1962, founding the Foundation for Research on the Nature of Man (FRNM). Rhine is often remembered as the man who "proved" that psychic powers exist.

Figure 123 Zener cards for ESP experiments.

In the classic Rhine experiment on ESP, the subject tries to guess or "call" the order of the five symbols when they are randomly placed in a table of 25 ESP cards. The likelihood of calling a card correctly by chance is 1:5 (one in five). Knowing this, it is possible to determine how often a particular score can occur by chance in a given number of calls. The Rhine assumption was that, when a subject obtains higher scores than could be expected, he displayed *extrachance* results or ESP. (An abnormally low score also denotes a paranormal skill, as it can be implicit in the subject's mind that he doesn't want to guess the correct card and therefore selects the wrong card.)

Today, in some sites on the Internet, you can find Zener card ESP tests where the cards are placed at random on your monitor and you make your choice while trying to sense which cards have been selected.

Lousia E. Rhine proposed the theory that ESP starts in the unconscious (a depository of memories, hopes, and fears). At this point, there exists a contact between the objective world and the center of the mind. The person remains unaware of this contact until or unless the information is brought to the conscious level. Carl G. Jung, in his time, proposed a similar theory to the effect that the conscious mind has subliminal psychic access to the collective unconscious, a vast repository of the cumulative wisdom and experiences of all humans.

Some criticisms exist today of the experiments made in the past. It is important for the reader to understand them, because they can be decisive in the experiments performed without projects. The criticisms are as follows:

1. *The "file drawer" effect.* In some cases, only favorable results have been published. When working with a large base of experimental data, the probability increases that some results will be omitted from the mean values. The experimenter is tempted to view some events as proof of ESP when they really are only the result of normal chance.
2. *The results are inconsistent or can't be repeated.* Another factor that the experimenter must watch for in ESP is preconceived or previously learned knowledge. This concerns any information that might influence the subject's perception. For instance, if a mother says that she senses that her son may suffer a fall when playing soccer on a specific day and time, it could be because her son has already had such an experience. Her sensation must be suspect, as it may be based on knowledge of the son's previous performance.

Other Terms

In the past century, ESP was called *cryptesthesia* and also *relesthesia*. Rhine was the first to use the term *general extrasensory perception (GESP)* to include other paranormal abilities such as telepathy and clairvoyance. So, as extensions of ESP, we can add other sensory paranormal abilities such as:

- clairvoyance
- clairaudience
- far-touch
- radiesthesia
- psychometry
- mind reading and telepathy

Clairaudience. This is the ability to hear paranormal information. It can be considered to be a form of ESP.

Psychometry. This is the gathering of information by touching physical things and objects. This also can be considered to be a form of ESP.

Precognition. This refers to the ability to see into the future. Because ESP doesn't limit the ability to sense images and sounds to any part of space-time, precognition is also considered to fit in that category.

Psychokinesis or Telekinesis (PK). This refers to the ability to move objects by focusing the mind on them. It is also called *far-touch*. Psychokinesis is a form of PSI, and it can be extended to other abilities than only moving objects. Bending metals and determining the outcome of events are also included as PK.

The term *psychokinesis* comes from the Greek words psyche, meaning *breath, life,* or *soul,* and kinene, meaning *to move.*

History

As with ESP, occurrences of psychokinesis phenomena have been recorded since ancient times. Among these occurrences, found in biblical and many other texts, we can find miraculous healings, luminosities, apports,* and other physical phenomena associated with holy persons and adepts of magic groups. The "Book of Acts" in the Bible describes an example of PK phenomena in the section where St. Paul and Silas, imprisoned in Ephesus, prayed and sang hymns to open the prison doors.

In the 19th century, D. D. Holmes was known for his ability to levitate and to handle hot coals without being burned. During that time, there were persons

---

* An *apport* is an object that materializes during a séance. Believers see apports as gifts or signals from spirits. When a medium makes an apport disappear, it is referred to as a *deport*.

known as *electric people* who experienced a *high-voltage syndrome*. Those persons made knives and forks cling to their skin, and with a touch they could send furniture flying across a room.

PK research has been a fast-growing area of interest since 1930. J. B. Rhine, working at Duke University (North Carolina) in 1934, was one of the first to conduct experiments in this field. He found that it was possible to influence the fall of dice, making them roll certain numbers or number combinations.

Rhine did not immediately publish his findings, for many reasons. One is that PK suffered a dubious reputation at that time, and the other was the experiments were very inadequately controlled. Later, Rhine divided PK in two categories: *macro-PK,* or observable events, and *micro-PK,* or weak or slight effects not observable by the naked eye.

In 1960, a new method of testing micro-PK was created by the American physicist Helmut Schmidt. He built an apparatus known as an "electronic coin flipper," which operated on the random decay of radioactive particles. As the decay rate is not affected by any physical quality such as temperature, pressure, magnetic field, etc., the rate of emission is completely unpredictable and cannot be manipulated by fraud. In the experiments, the subjects were invited to exert their mental energy to influence the flipping of the coins. The number of heads and tails was indicated by lamps.

Schmidt also studied animal-PK, finding some interesting results. However, the interpretation of the results was difficult, as he theorized that the experimenter could influence the results by using his own PK on the experimental subjects.

One of the most notable macro-PK events was what is now called the Geller effect. In the 1960s, the psychic Israeli, Uri Geller, amazed television audiences with his metal-bending feats. But Geller was unable to duplicate the feats under laboratory conditions.

Today, many researchers work with PK using sophisticated methodology. The experiments are focused on psychics, mediums, and other people who can apparently influence objects and materials.

One present study is under way with Ingo Swann, a New York artist and psychic who can change the temperature of a nearby object by one degree and also can affect the magnetic field detected by a magnetometer.

Other Types of PK

Other types of PK have been studied but are viewed with a fair amount of skepticism. One of these is the poltergeist activity. Such activity includes unexplained repeated sounds, breaking of china, flying rocks, movement of heavy furniture, and other mysterious manifestations in a small area.

Another type is thought by those who experienced it to be associated with death, danger, or emotional crisis. These are cases in which people report clocks that stop, falling objects (pictures from the wall are most commonly reported), and shattering of glass objects. Many people believe that these incidents indicate death or an accident involving relatives or loved ones. Experiments also are being conducted to determine the existence of a "retro-PK" where the subjects can influence an event in the future or in the past!

PK is not generally accepted by scientists, but many parapsychologists believe that well controlled experiments can establish its existence.

PSI. In 1946, British psychologists Drs. Robert Thouless and W. P. Weisner proposed the word *PSI* to designate both extrasensory perception (ESP) and psychokinesis (PK). PSI is the twenty-third letter of the Greek alphabet and is commonly used in parapsychology to include both phenomena of PK and ESP because they are closely related.

Theories concerning the functioning of PSI are very difficult to formulate, because it defies laboratory experiments. Researchers have not been successful in describing its activity in terms of physical sciences.

Experiments with PSI involve the measurement of the involuntary physiological processes in the autonomic nervous systems of test subjects. The most common measurements are *galvanic skin response (GSR),* which records the activity of the sweat glands; *plethysmography,* which measures the changes in blood volume in the fingers; and *electroencephalography (EEG),* which measures brain activity.

Psychography. This refers to the ability to write messages from beyond. This paranormal ability of a subject can be considered to be a type of ESP. Without the use of the normal senses, the person can pick up messages from the beyond and transfer them directly to a piece of paper. The person writes messages "automatically," normally in a trance, without any knowledge of their content.

Mind Reading and Telepathy. If a subject is able to probe the mind of others and "read" their thoughts as if their minds were an open book, we have the phenomenon called *mind reading*. We consider this paranormal ability as a form of ESP, as it is a paranormal sense used to pick up information from someone's brain. Telepathy is different: if the subject can mentally send some kind of information to another person, the phenomenon is called *telepathy*.

Reaver. This refers to the power of an individual to excite or slow the molecular motion of atoms. A person with this paranormal ability is able to suddenly or slowly cool the immediate surroundings, set someone's hair on fire, freeze someone's body, or extinguish a candle flame. It is a form of telekinesis.

Radiesthesia. This is someone's ability to divine or dowse, through indicators such as rods and pendulums. This method embraces much more than just the discovery of water, treasures, and metals. Radiesthesia or *radiesthesie* (the French word) has been used in discovering missing persons and in performing medical diagnoses.

Radiesthesia starts from the idea that either a rod or a pendulum amplifies the person's sensitivity. Usually, the pendulums consist of small balls or cones attached to the end of a stick via a thin string, as shown in Fig. 124.

The string can be (preferably) nylon. It is necessary that the person be experienced in using the pendulum. We want no movement of the "bob" except what is caused by the influence of the sought-after object. For further explanation about the use of the pendulum, we suggest that the reader consult related literature.

History. The term *radiesthesie* was coined in 1930 by Abbe Bouly in France, where the rod gave way to the small pendulum as an indicator. In 1933, the British Society of Dowsers was founded.

Transcendental Meditation (TM). Transcendental meditation is a system by which a person can achieve or reach a *fourth state* of consciousness, also called *transcendental consciousness.*

History. TM is cited in the Vedas sacred writings going back to 1000 BC. Over the centuries, it has been transmitted by such men as the eighth-century Hindu philosopher Shankara, and in the twentieth century by Guru Dev (meaning *divine teacher*), who taught Maharishi Mahesh Yogi.

Maharishi Mahesh Yogi, after graduating in physics at Allahabad University, spent two years in a Himalayan retreat and then began teaching about TM in India. After that, he traveled throughout the East and West, training teachers to spread TM and the *science of creative intelligence,* whose end is to integrate all knowledge.

Physiological changes occur in TM practitioners. These changes include the lowering of respiration, heart rate, blood pressure, and lactase (a chemical in the blood associated with strenuous activity and stress).

3.2.2 Our Practical End

We could devote the following pages to describing all of those paranormal phenomena and discussing whether they are valid, or we could try to explain them based on the modern science. But that is not the aim of this book. It is up to the

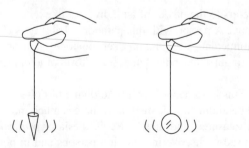

Figure 124 Experimenting with a pendulum.

reader to find information about the phenomena, which is available from many specialized sources. (There are thousands of them on the Internet; the reader has only to type the appropriate keywords into some search engines such as Alta Vista, Yahoo, Infoseek, Lycos, and others.) The aim of this book is to provide the reader with electronic devices that will improve some practical experiments.

As the preceding descriptions of paranormal phenomena show, all are associated with changes in physical quantities in our body. Many of them are visible and easily detectable (macro changes), but others need the aid of some device (micro changes) to be recognized. This means that the reader, who can now extend his senses with the aid of electronic technology to detect the micro changes, is much closer to new discoveries than are individuals who are not so equipped.

We can see only a small part of the electromagnetic spectrum, and we can hear only a very narrow band of the sound spectrum. At the same time, we can't feel electric signals or small temperature changes in a body without touching it (and sometimes even if we do touch it). To remedy this situation, the following pages provide the reader with modern electronic technology that can extend our senses or stimulate them to produce the paranormal phenomena.

As in previous parts of this book, the projects are simple but sensitive and can be useful for all readers who are experienced in building electronic devices. Using cheap and easy-to-find parts, the reader can extend these experiments, advancing many steps in the direction of the frontier beyond.

The Circuits

The important point to mention is that the circuits described in the next sections can be used for several purposes. For instance, a biofeedback circuit that uses a temperature sensor to detect and control changes in skin temperature can be used to control the temperature changes of the same person in an ESP or telekinesis experiment. Other applications of the same circuit include experiments in radiesthesia, which involves detecting changes in some physical quantity identified by a sensor in the presence of water or minerals.

For convenience, we can divide the circuits to be described into three groups.

1. *Training Devices*

   In some cases, the manifestation of some paranormal phenomena is not natural. This means that the individuals must be induced into the paranormal state. Induction can be important to find paranormal abilities in persons who, in normal life, are not aware of them.

   Many procedures are adopted to put the individual into the necessary mental or physical condition to create paranormal phenomena. In many cases, it is possible to achieve the necessary conditions only after a certain training process. The adepts of transcendental meditation, for instance, repeat words and sounds over and over again to achieve the proper mental state. Biofeedback is often used to train the individual in creating the necessary mental state for a paranormal manifestation. Hypnosis induction devices can also help the individuals to achieve semiconscious, trance states in which the paranormal abilities can appear.

2. *Monitoring*

    When working with paranormal phenomena, it is not easy to see if an individual is making progress toward achieving the necessary mind (or body) state to proceed with the experiment. As we said, we can't see temperature changes of the individual body, skin resistance changes, or blood pressure alterations. Devices that can extend our perception help with the research to make the experiments easier. The same type of device can be used to detect temperature changes or magnetic field changes in a body during a PK experiment.

3. *Induction Devices*

    Word repetition, flooding an ambient with light, or the creation of some form of "energy" can induce individuals to exhibit paranormal skills. There are many ways to use electronic devices to achieve this, by filling ambients or exciting senses using electronic devices and circuits. Stroboscopic lights, special sound generators (tone, ultrasonic, and noise), and skin and nerve stimulators are some examples of these devices.

3.3 Sony's Seven-Year Paranormal Research Effort

Over a period of seven years, starting in 1989, Japan's enormously successful electronics giant, the Sony Corporation, worked in paranormal research. For this task, they created the Extrasensory Perception and Excitation Research (ESPER) laboratory. Sony shied away from discussing the lab and preferred to avoid publicity about its existence and what was going on inside. There was no mention of it in Sony's annual reports or other official literature.

ESPER lab was used for studies of qi,[*] presentiment (prevision or precognition), synchronicity, mind-body interactions, consciousness, the sixth sense, and supernatural phenomena. The founder was Yoichiro Sako who, in 1989, approached one of Sony's two founding fathers, Masaru Ibuka, about starting a special department to study qi.

In 1988, Ibuka established the Pulse Graph Research Department to work on a device that was claimed to identify health problems by measuring the pulse. In 1990, he extended the research into measuring other physiological parameters that could change in the body while qi masters tried to alter patients' qi energy. Skin temperature was one of these additional measurement parameters. One year later, in 1991, Sako had convinced Sony's founders to establish a separate laboratory where his studies could be continued, including research into other paranormal phenomena such as psi. Another important event occurred in 1995, when the ESPER lab was split from the Research Institute of Wisdom and became part of the company's Research and Development Division.

---

[*] The term is pronounced "ch'i" and in Chinese means *human science* or *bioenergy*.

Some years ago, Sako was interviewed in Las Vegas, and the conversation was published by the magazine *Fortean Times* (vol. 115). According to Sako, the greatest success of the ESPER lab was in the field of clairvoyance. Experiments made with a young school girl revealed fantastic results. But the interesting part is that, when the interviewer asked about the techonology, the answer was surprising. Instead of describing a highly advanced technical process that is not easy to reproduce in a common laboratory, Sako answered simply, "It's low technology. High tech is not necessary."

According to the Fortean Times interview, a Sony spokesman named Masanobu told the *South China Morning Post*'s Benjamin Fulford, for a story that appeared some days afterward, "We found out experimentally that ESP exists." In the same interview, when asked by a reporter from *Fortean Times* if Sony had any psi-based products in works, he flashed a great big friendly smile and, after a long delay and a vigorous shake of the head, said, "There are no products. Not yet." Recently, Sony announced that the ESPER lab will be closing its doors "officially."

What is behind all these facts? Why is Sony closing the lab's doors? Did they discover something so important that it would be inconvenient to let the world know about its existence? Must further investigations be kept secret?

According to Bill Higgins, who worked for a company in New Jersey that attempted to create products based in PSI technology, "Sako was onto something." The words *derma sight* and *touch sight* seem to have a special meaning for Sony's researchers.

Some interesting questions arise from all of this. Why is an electronics company so interested in paranormal experiments? If the electronics are so important, is it necessary to use high-tech equipment in this field? Or was Sako correct that only low-tech approaches are required, meaning that simple circuits and devices can be used to obtain good results?

These questions can serve as an important starting point for readers who want to perform experiments using the projects that are presented next. We can infer that much more can be discovered than you may be imagining. The "aura" of mystery characteristic of the oriental culture, when added to all these contradictory indications, can be a source of much more enthusiasm for the reader who wants to conduct practical experiments.

3.4 Biofeedback Experiments and Projects

> *It isn't only art that's incompatible with happiness;*
> *it is also science. Science is dangerous; we have to*
> *keep it most carefully chained and muzzled.*
>
> Aldous Huxley (1894-1965)

Biofeedback is a technique by which persons are trained to make certain physiological changes in their bodies using signals from within. Physical therapists use it to help stroke victims recover movement in paralyzed muscles. Physiologists use it to help tense and anxious patients learn to relax. Using biofeedback, patients can control blood pressure, skeleto-muscular responses, heart rate, peripheral skin temperature, and bowel functions.

But it is not only in medicine that biofeedback can be used. In paranormal experiments, biofeedback can help the researcher to find some special mental or physiological state that is favorable to the experiments as in PSI, clairvoyance, psychometry, telepathy, and transcendental meditation.

Biofeedback means to sense some kind of information about the person's body and use it to control or change some physiological process. For instance, variations of one's blood pressure can be fed back in the form of information that can be sensed by our sense organs (the flash of a lamp or the pitch of a tone) so that the person becomes aware of those blood pressure changes. With this awareness, the subject then has a chance of assuming control over that biological function. Hence, the term *biofeedback*: by monitoring various bodily functions, people can control their operation (*bio* meaning *life* in Greek).

Electronics Is Not Necessary but Interesting

It is not necessary to enlist the aid of electronic devices to sense changes in our body. Adepts of transcendental meditation and other mystic practices can use their own senses to take control of their physiological functions. For example, the sound of the breath or the beat of the heart can be used by trained adepts.

The continual repetition of special sets of words or sounds used by adepts of transcendental meditation can be considered a form of biofeedback, where the sounds are the information to be picked up by the ear and used to control their physiological and mind functions to achieve the relaxed state.

Where Electronics Can Help

For the researcher who is experienced with electronics, some circuits can increase the chance of improved results in the paranormal experiments. The use of electronic equipment can also help to identify individuals with unknown paranormal abilities and to open unexpected fields of research through the creation of new experiments.

As with the other paranormal experiments described in this book, there exist many simple circuits that anyone who has some experience with electronic construction can assemble in few hours using low-cost parts. To help the paranormal researcher who wants to try some experiments in this field, we start our projects to address paranormal sensing with some simple biofeedback circuits.

It is important to mention that many of the projects described herein, now suggested for applications in biofeedback experiments, can also be used in other paranormal experiments and even in other scientific fields. Some biofeedback circuits can be used simply to correct disorders such as tension and migraine headaches, hypertension, muscle tension, and general relaxation.

Experimenting with Paranormal Skills

Warning:
The devices described here are indicated for experimental purposes only. Do not try to use them as a solution for physical disorders without first consulting a medical doctor.

3.4.1 How a Biofeedback Circuit Works

The simplest circuit for a biofeedback device is shown as a block diagram in Fig. 125. Some form of sensor picks up a signal from a person's body and applies it to an amplifier stage.

The sensor can work with any physiological state that is dependent on changes in bodily functions, such as

- Temperature
- Blood pressure
- Skin resistance
- Muscular tension
- Heart beat
- Breath
- Cerebral waves

The amplifier configuration depends on the signal level produced by the sensor. With a sensitive sensor, such as a diode or a light-dependent resistor (LDR), a single transistor is enough to provide a good amplification of the signal. Using less sensitive sensors, or trying to detect very small changes in the person's body, an operational amplifier (opamp) is needed.

The signal picked up from the output of the amplifier stage is used to control some kind of signal generator circuit. This block produces some kind of stimulus in the person, providing sensory feedback of the detected physical changes.

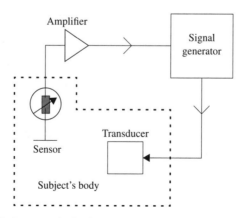

Figure 125 Block diagram of a biofeedback device.

An audio oscillator, for instance, can produce a tone with a pitch that changes when the signal produced by the sensor changes. In this case, we have *auditory feedback*. A light flasher can be used to produce *visual feedback*.

Another possible configuration is a high-voltage generator that is used to stimulate the person to a degree that depends on the level of the signal produced by the sensor. Figure 126 shows how this *electro-shock feedback* works.

In a complex biofeedback system, many sensors can be used to pick up signals from different parts of a person's body, and the feedback stimulus can be simultaneously visual, auditory, and sensory. The circuits described in this book use all of these stimulus types and can be adapted to form complex research systems.

3.4.2 The Paranormal Aspect

The number of paranormal experiments that can be created using a biofeedback circuit are unlimited. Some applications of paranormal circuits and devices are

1. *ESP.* A biofeedback circuit can be used to help the subjects of the experiment to achieve a mental and physiological state that is favorable to the experiments. Concentration, relaxation, trance, fourth state of consciousness, and other mind states needed to perform the experiments can be achieved more easily with the aid of biofeedback circuits. The researcher can also investigate how biofeedback can influence the results of an experiment. Another application is for monitoring the state of the subject of the experiment.
2. *Clairvoyance and psychometry.* A biofeedback circuit can be used to help the subject to find the ideal mind state for the experiments. Changes in body functions can be monitored with biofeedback circuits.
3. *Psychokinesis.* Here also, biofeedback can help the subject of the experiments to achieve special mind states before the sessions or to monitor the subject's PK abilities.

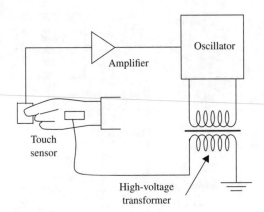

Figure 126 Electro-shock biofeedback.

Experimenting with Paranormal Skills 173

4. *EIP and EVP.* It can be interesting to see how biofeedback can help the researcher to achieve special mind states during the experiments, and how this can influence the nature of the received voices and images.
5. *Auras.* A biofeedback circuit can be used to induce the "patient" into special mind states before taking the picture of his aura.

3.5 Experiments

Again, the number of experiments that the reader can create using the described electronic circuits is unlimited. Many configurations of these circuits and devices can be conceived to test paranormal abilities. Again, warn the reader against being led into false interpretations of the results when making the experiments. (See the recommendations in Parts 1 and 2 about sounds and images, e.g., Section 1.10.)

Project 25: Temperature Change Monitor

The first circuit of the series is very simple and can be built from a few cheap, easy-to-find parts. This simple temperature change monitor can detect variations of temperature in any small part of your body (your fingers, for instance) or other objects, allowing you to use this information to control some physiological function or simply monitor temperature changes in experiments.

The circuit is powered from two AA cells, and the current drain is very low, providing battery life of several weeks.

3.5.1 Experiments

Monitoring the temperature changes in living beings or inanimate objects can be important in many paranormal experiments.

- Putting the sensor between your fingers, you can detect temperature changes by the movement of a meter's needle. In a biofeedback experiment, you can then act in some manner to control it or make it move in the opposite direction. The sensor can be affixed to other parts of the body as appropriate for the intended experiment.
- Changes in the temperature of an object can also be monitored for experiments in PK and ESP. Use the sensor to detect if there are any temperature changes in an object you're trying to move with your mind. You can also monitor temperature changes in an object that is observed or used as the target in an ESP experiment. Place the sensor on a plant's leaf to detect temperature changes in experiments involving PK and ESP.
- In radiesthesia, the circuit can be used to detect small temperature changes in an object used as a sensor. This can be induced by the presence of water or minerals or due to the pendulum action.

- Reaver is another area of experimentation in which this project can be useful. Temperature changes in the material to be bent can be detected during the experiments.
- In transcendental meditation experiments, the sensor can detect when the skin temperature of a subject falls, indicating that the fourth state of consciousness or trance has been achieved.
- Other paranormal fields include ghost or spirit manifestations altering the temperature of subjects (spontaneous combustion).

An important point to observe about this project is that the reduced thermal capacity associated with the small size of the sensor makes the circuit respond very quickly to temperature changes.

How It Works

When a silicon diode is reverse-biased, the current flow or resistance depends on the junction temperature. More charge carriers are liberated in the junction when the temperature increases, reducing the resistance and increasing the current. Figure 127 shows how the resistance of a reverse-biased diode changes with temperature. This property of a silicon diode (or any other device using silicon junctions) allows it to be used as a sensitive temperature sensor.

In this project, a reversed-biased silicon diode is wired to the base of a silicon transistor acting as an amplifier. The transistor, a resistor, and a potentiometer form a Wheatstone bridge having a current meter as a zero indicator.

As the equilibrium of the bridge depends on the temperature, the potentiometer can be used to zero it at some reference temperature, and the meter will indicate any change.

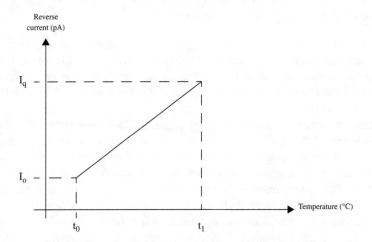

Figure 127 Reverse current increases with temperature in a diode.

Assembly

Figure 128 shows the complete diagram of the thermal biofeedback device. As the circuit is very simple, not critical, and appropriate for a beginner, a terminal strip can be used as the chassis as shown in Fig. 129.

Any general-purpose silicon diode, such as the 1N914 or 1N4148, can be used as a sensor. To protect the sensor's terminals from physical contact, which can alter the reading and current indicated by the meter, isolation must be provided as shown in Fig. 130a. This isolation consists of two pieces of plastic tubing to cover the common wires soldered to the diode terminals.

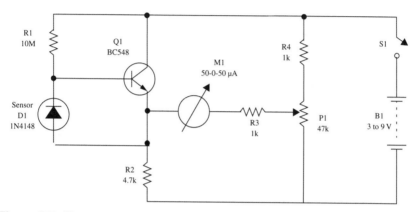

Figure 128 Temperature change monitor.

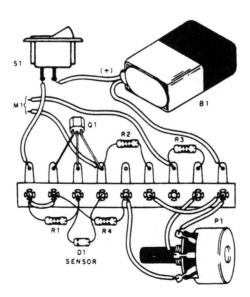

Figure 129 Temperature change monitor mounted using a terminal strip.

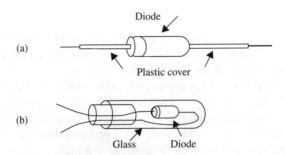

Figure 130 Protecting the diode from moisture.

Another form of isolation consists of placing the sensor inside a glass tube or encasing it in an epoxy block (Fig. 130b). The proper form of protection for the sensor depends on the experiment. Remember that, in the case of the glass tube, the heat flows very slowly through it, resulting in a slow response.

For M1, any microampere current meter can be used. You can get best results using a meter with a 50 µA range and zero in the center of the scale.

The circuit is powered from two or four AA cells placed in a holder.

Using the Circuit

Turn S1 on and adjust P1 to put the needle of the indicator at the center of its scale. For meters with zero in the center, this means a zero current flowing through the circuit. As we can detect only temperature changes, we can use other types of meters by setting the needle to the center of the scale, meaning that some amount of current is flowing through the circuit, and it is possible to detect current changes.

Place the sensor between your fingers and observe the change in position of the meter's needle. If the needle tends to go to the left (indicating a temperature drop), reverse the meter's connections.

Wait until the needle stops in a fixed position. Now you are ready to start with your experiments. Without changing your fingers' pressure on the sensor, try to change the position of the needle by controlling the temperature of your body.

Suggestions

- Sensitivity is regulated by R1 and Q1. You can increase the sensitivity of the circuit in two ways: by replacing R1 with a larger resistor (22 to 47 MΩ) or by replacing Q1 with a Darlington transistor such as the BC517.
- Two sensors can be used at the same time. An additional diode is placed between the transistor's base and the ground line of the circuit. This arrangement allows the researcher to simultaneously control the temperature of the fingers of the right and left hand, trying to alter it in a differential mode.

Parts List: Project 25

Semiconductors

Q1   BC548 or equivalent—any general-purpose silicon NPN transistor

D1   1N4148, 1N914 or any silicon diode

Resistors

R1   10 kΩ, 1/8 W, 5%—brown, black, blue

R2   4.7 kΩ, 1/8 W, 5%—yellow, violet, red

R3, R4   1 kΩ, 1/8 W, 5%—brown, black, red

Miscellaneous

P1   47 kΩ potentiometer

M1   Microampere meter, 0–50 to 0–200 μA single or with zero in the center of the scale

S1   SPST, toggle or slide switch

B1   3 to 9 V, AA cells or battery

Terminal strip, cell holder or battery clip, wires, solder, plastic box, etc.

Project 26: Temperature-Controlled Oscillator

Watching the movement of a needle to control physiological functions is one form of visual feedback, as shown in Project 25. In this project, we provide an auditory project wherein information about temperature changes in our body is provided by alterations in an audio tone.

To be useful as a feedback circuit, the "patient" must be able to control the tone produced by the device by changing his body temperature. But other experiments can be performed with this device such as described for the previous project. The use of a tone can free the person who is training or being observed to close his eyes and use only his ears to feel the temperature changes.

3.5.2 Experiments

The basic application for this circuit is as a training device or as biofeedback, helping the operator to achieve special mental or physiological states. The person must maintain the tone at a predetermined frequency or try to change it by modifying his physiological state.

As a monitoring device, the circuit can detect changes in the temperature of a human body or an object in PK experiments, as described in the previous project.

All of the experiments performed in the previous project can be repeated using this circuit.

How It Works

A general-purpose silicon diode is used as a temperature sensor, as in Project 25. The current flowing from the sensor is amplified by a transistor (Q1) and used to bias a two-transistor audio oscillator. In this oscillator, feedback that keeps the circuit in operation is provided by C1. This capacitor, with the resistance represented by Q1, determines the operational frequency of the circuit.

Considering that the resistance of Q1 changes according the current biasing its base, which is determined by the temperature of the sensor, the frequency of the tone also is dependent on the sensor's temperature.

R1 is used to adjust the sensitivity of the circuit and the central frequency of the tone in an experiment. The circuit is powered by two or four AA cells and uses common parts.

Assembly

Figure 131 shows the schematic diagram of the temperature-controlled oscillator. Like the previous project, this is also very simple and appropriate for the reader who is less experienced with electronic assembly, and it uses a terminal strip for mounting the components (Fig. 132).

Keep the wires and component terminals short to avoid shorts. The wire to the sensor can be as long as 40 cm. Avoid the use of longer wires, which can pick up noise that affects the circuit operation.

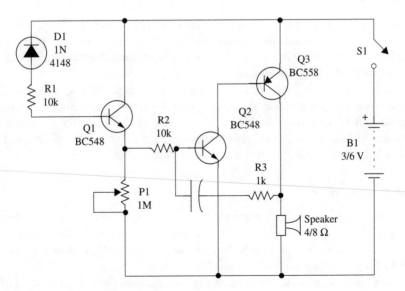

Figure 131 Temperature-controlled oscillator.

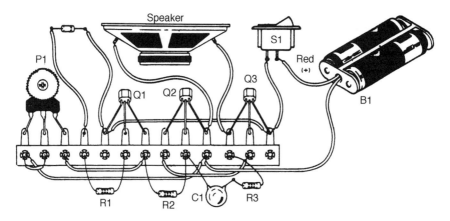

Figure 132 Terminal strip used as the chassis.

Any general-purpose silicon diode can be used as the sensor. Protect the terminals with small pieces of plastic tubing as in the previous project or according to the experiments you have in mind.

A plastic box can be used to house the circuit. Dimensions are determined basically by the size of the loudspeaker. Any small loudspeaker with a diameter between 5 and 10 cm can be used in this project.

Using the Circuit

Turn on S1 and adjust P1 to any low tone from the loudspeaker. Then, place the sensor between your fingers and wait a few seconds. The tone will change, increasing in frequency.

Find the correct adjustment for P1 to get the widest tone variation when you place the sensor between your fingers. Experiments can be conducted as described in Project 25.

Suggestions

- Replace the transistor Q1 with a Darlington, such as the BC517, to increase sensitivity. With a high-gain transistor, the circuit can detect smaller changes in body temperature.
- C1 determines the frequency of the tone. Replace it with capacitors rated between 0.022 and 0.22 µF and note the difference.
- We can generate a pulsed audio tone that varies with temperature by replacing C1 with various capacitors rated between 0.47 and 1 µF.
- Replace Q1 with a TIP32 if you want to power the circuit from 12 V power supplies. Place the TIP32 on a heatsink and use a 10 cm × 5 W or larger loudspeaker.

Parts List: Project 26

Semiconductors

Q1,   BC548 or equivalent—any general-purpose NPN
Q2   silicon transistor

Q2   BC558 or equivalent—any general-purpose PNP silicon transistor

D1   1N914, 1N4148 or any general-purpose silicon diode

Resistors

R1,   10 kΩ, 1/8 W, 5%—brown, black, red
R2

Capacitors

C1   0.047 µF—metal film or ceramic

Miscellaneous

P1   1 MΩ—trimmer potentiometer

S1   SPST—toggle or slide switch

SPKR   4 or 8 Ω—5 to 10 cm—loudspeaker

B1   3 to 6 V—two or four AA cells

Terminal strip, cell holder, plastic box, wires, solder, etc.

Project 27: Light and Dark Controlled Oscillator

The amount of light (e.g., controlled by the shadow of your hand) falling on the light-dependent resistor (LDR) in this project will determine the tone produced by the circuit. Extending your hand and keeping it over the LDR, you can control the tone produced by this circuit by letting more or less light fall on it. This is one of the applications for using this circuit as a feedback device.

This form of biofeedback can be used to help a person to concentrate his mind to keep control over the arm muscles for meditation. Other paranormal experiments are described below.

3.5.3 Experiments

- Place the subject's hand on the sensor, and use a light source to illuminate it as shown in Fig. 133. The hand isn't totally opaque, so some light can pass through it. The circuit can be adjusted to produce a tone from this light. Alterations in this tone indicate some modification in the blood flow in the hand, af-

Figure 133 The hand is not completely opaque, and some light can pass through it.

fecting its opacity. Experiments with trance conditions and transcendental meditation, or during ESP sessions, can incorporate the detection of this physiological parameter using this circuit.
- Place transparent and semi-transparent solid objects on the sensor and see if, using your mental power, you can alter the amount of light passing through them. This way, you can perform experiments involving PK.
- A mirror at the end of a nylon wire and placed inside a transparent glass chamber can be used to detect paranormal skills as shown in Fig. 134. The person who is the subject of a PK experiment must move the mirror using his mind. A laser beam focused on the mirror and reflected to a screen forms a very sensitive movement detector. Very small changes in the position of the mirror will produce a large displacement in the laser beam focused on the screen.

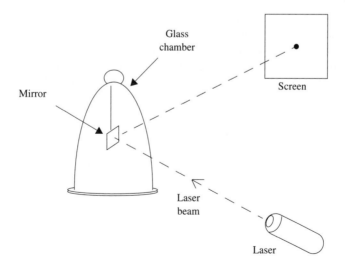

Figure 134 Detecting micro-movements of a mirror in a PK experiment.

- Repeat the above experiment but place the mirror on a pendulum.
- Ghostly and other paranormal phenomena can be detected by changes in the brightness of ambient light, which are translated into changes in the tone generated by this circuit. Small changes in the lamp brightness due to "paranormal" phenomena that are not visible to the naked eye can be more easily detected by the change in the tone. A candle can be used as an intermediate source of light, as shown in Fig. 135, where the tone changes detected by the circuit can be used as feedback or to detect some paranormal effect. The circuit is more sensitive than the human eye and can even detect radiation in the IR (infrared) and the UV (ultraviolet) spectrum, extending its application to many experiments involving "invisible" light sources. Use a filter to perform experiments with specific light wavelengths.

Power for the circuit comes from two or four AA cells, and the unit can be placed inside a small plastic box. The reduced dimensions of this device allow it to be used in virtually any location.

How It Works

As is the case with many other circuits described in this book, this configuration is very simple. Two complementary transistors form a low-frequency oscillator. The frequency is determined by C1 and the biasing network formed by the LDR, P1, and R1.

The LDR or a CdS cell is the light sensor in this circuit. When the sensitive surface of this device is exposed to light, its resistance changes.

As shown by Fig. 136, the resistance of an LDR decreases when the amount of light falling on it increases. In the dark, this device shows a resistance of many megohms, but this resistance falls to some thousands or even hundreds of ohms when illuminated by sunlight.

The LDR is used as a tone control in this circuit. Adjustments can be made via P1, setting the circuit to its oscillation limit by a particular amount of light. Therefore, by controlling the amount of light that falls on the LDR, we are able to control the frequency of the oscillator within a wide frequency range.

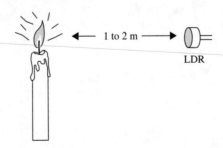

Figure 135 Using a candle as a light source in paranormal experiments.

Experimenting with Paranormal Skills   183

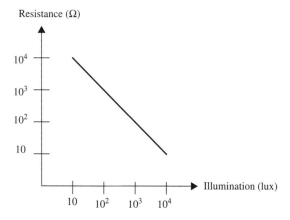

Figure 136   Resistance as function of illumination in an LDR.

Many components can be substituted in this circuit to alter its performance, as we will see in the suggestions.

S1 is used to cut the oscillation any time. You can also use a reed switch and operate it with a magnet, or use another kind of switch.

Assembly

Figure 137 shows the complete diagram of the light/dark controlled oscillator. Because the circuit is very simple and not critical, it can be mounted on a terminal strip as shown in Fig. 138.

Keep the wires short to avoid introducing instabilities in the circuit. Any LDR can be used in this project. Types with diameter of 1 cm are suitable. You can increase the sensitivity by installing the LDR inside a cardboard tube and placing in front of it a convergent lens.

All the components can be installed in a plastic box. The size of the box depends basically on the loudspeaker's diameter. Types with diameters from 5 to 10 cm are indicated for this application.

A power switch was omitted in this project, as the circuit can be turned off by removing the cells from the holder.

Using the Circuit

Turn on the power supply and adjust P1 until you obtain a tone with the ambient light. Then place your hand in front of the LDR and see how its shadow can alter the tone frequency.

Biofeedback experiments can be performed by placing the circuit in any ambient and adjusting P1 to any tone. Then, placing your hands in front of the LDR, try not to maintain a steady tone.

184 Part 3

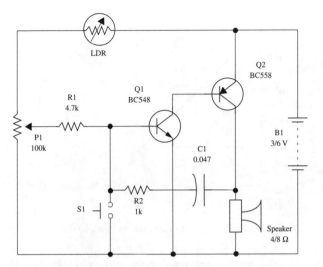

Figure 137 Light and dark controlled oscillator.

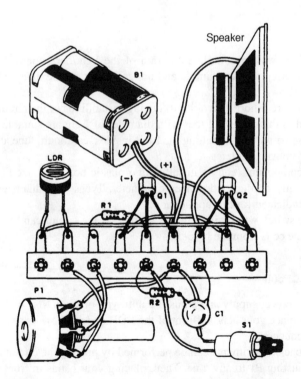

Figure 138 The project is mounted using a terminal strip as chassis.

Suggestions

- Replace C1 with an 0.47 to 1 µF capacitor and adjust P1 to obtain spaced pulses. You will then have a feedback metronome, and you can conduct some experiments using it. In transcendental meditation and biofeedback experiments, you can adjust the metronome's pulse rate to be the same as your heartbeat or your breath rhythm.
- Place a color filter in front of the LDR to work with colored light sources. This way, the circuit will be able to detect light changes only in a specific color.
- A second LDR can replace S1 in an experiment involving two light sources.
- The circuit can be used to detect changes in an aura if the LDR is placed inside a cardboard tube with a convergent lens. Point the LDR at the aura and see how the tone changes.
- The loudspeaker can be replaced with a headphone for some experiments.
- Place the LDR inside a cardboard tube with a convergent lens in front of it to add directivity and sensitivity to the project.

Parts List: Project 27

Semiconductors

Q1   BC548 or equivalent, any general-purpose silicon NPN transistor

Q2   BC558 or equivalent, any general-purpose silicon PNP transistor

Resistors

R1   4.7 kΩ, 1/8 W, 5%—yellow, violet, red

R2   1 kΩ, 1/8 W, 5%—brown, black, red

Capacitors

C1   0.047 µF—ceramic or metal film

Miscellaneous

P1   100 kΩ—potentiometer

B1   3 or 6 V, two or four AA cells

LDR Any LDR or CdS cell (see text)

S1   Pushbutton, normally open

SPKR 4 or 8 Ω × 5 or 10 cm loudspeaker

Terminal strip, cell holder, wires, plastic knob for P1, plastic box, solder, etc.

Project 28: Polygraph

When someone is interrogated, small changes in skin resistance can be associated with an anxious state of mind that results from telling a lie. This is the operating principle of the lie detector or polygraph.

But a circuit that can detect small changes of skin resistances can also be used in paranormal experiments. In fact, special mental states due to paranormal phenomena (or due to a state of concentration, trance, or the "fourth state of consciousness"), if detected, can be used to improve your experiments. The circuit described here is very simple and sensitive and can be used as a lie detector or to detect skin resistance changes for any reason.

3.5.4 Experiments

Some proposed experiments using this circuit include the following:

- Try using the skin resistance data from the current meter to perform biofeedback experiments.
- Skin resistance can be associated with physiological and mental states in ESP and PK experiments. Changes in skin resistance during ESP trials can be detected and linked to results or mental states of the individuals under examination. In telekinesis and psychometry studies, the researcher can detect whether any change in the skin resistance of a subject occurs during the experiments.
- The resistance of plants and organic solutions can be monitored with this circuit, detecting the influence of paranormal forces during experiments. Using your mind, try to alter the resistance of a water/vinegar mixture in a cup. As electrodes, use two pieces of bare wire, 10 to 12 cm long.
- Attach the electrodes to a plant leaf and see if paranormal phenomena affect their resistance. Experiments in PK and ESP can be conducted using plants as sensors. (Clever Backster discovered that plants have some paranormal abilities by monitoring them with a polygraph. See the book, *The Secret Life of Plants,* by Peter Tompkins and Christopher Bird.)
- Experiments involving pendulum action and changes in the skin resistance of a subject can be performed. Changes can indicate some alteration in physiological or mental states that is unseen by the naked eye or other detection processes.
- The circuit can be used to detect many other paranormal phenomena based on the electrical resistance changes of living beings and things. The skin resistance of "sensitive" subjects can change during paranormal manifestation. This applies to the resistance of living things such as plants and live solutions (lacto bacillus, bacteria colonies, etc.).

How It Works

Two transistors in a Darlington configuration form a dc amplifier. The current is drawn from the two electrodes placed on the skin of a person. Any change in skin resistance will produce a variation in the resistance between the collector and emitter of Q2. Q2, R2, R3, R4, and P1 are wired to form a Wheatstone bridge. The zero (null) detector in this bridge is a sensitive current meter.

By placing the electrodes on the skin of a person, it is possible to zero the bridge using P1. Any change in skin resistance will be transposed into a strong change in the equilibrium of the bridge, causing current to flow through the meter.

Assembly

Figure 139 shows the diagram of the lie detector or polygraph. As the circuit is very simple, a terminal strip can be used as the chassis. Figure 140 shows how the components are placed on the terminal strip.

As a null indicator, you can use any current meter that can measure currents in the range of several microamperes to several milliamperes. An analog multimeter in the lowest current scale can be used for this task. Alternatively, you can use any microamperimeter with full-scale currents of 100 μA to 1 mA.

All the components can be housed in a small plastic or wooden box. Using a 9 V battery allows the use of smaller boxes as compared to using AA cells for the power supply. The on/off switch has been omitted. Place the cells into the holder to turn the unit on.

The electrodes used will depend on the application. Using the circuit as a skin resistance monitor or polygraph, the electrodes can be two metal rods. The rods have a length of 10 cm and a diameter of 1 or 2 cm. The subject must keep the electrodes in his hands, using constant pressure.

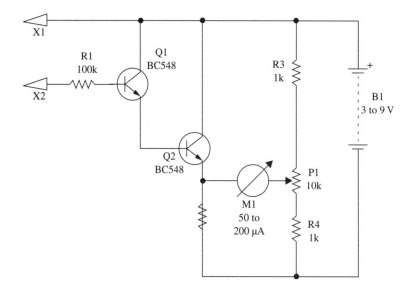

Figure 139 Schematic of the polygraph.

188   Part 3

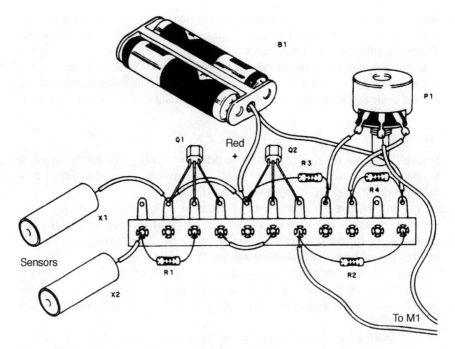

Figure 140 Circuit mounting using a terminal strip.

Another option for the electrodes is to use two small metal plates where the finger or the hand is placed, as shown in Fig. 141. Other electrode configurations are covered in the suggestions section as appropriate for the experiments.

Using the Circuit

Place the cells in the holder or the battery clip onto the battery terminals. Place the electrodes in your hand (without letting one touch the other) and adjust P1 to

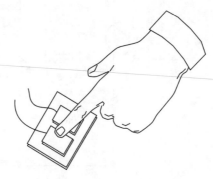

Figure 141 Two metal plates used as a sensor.

show a center scale deflection. Do not alter the pressure on the electrodes, and observe the needle indication on the meter. Any change in skin resistance will be shown by a movement of this needle.

In experiments involving plants, the electrodes can be two small metal plates fixed to a leaf by a clip. Don't use a clip that produces excessive pressure, as it can damage the leaf and alter the results of the experiments.

Suggestions

- You can monitor the bioactivity of plants by placing two electrodes on a plant leaf as shown in Fig. 142. Experiments involving telekinesis or the Backster effect can be performed using this configuration. Try to change the plant's physiological activity (detected by this device) by concentrating your mind on it or using it as a target in ESP or radiesthesia experiments.
- In plant experiments, the electrodes can be two pieces of rigid wire placed in the flower pot with the plant.
- The two transistors can be replaced by one low-power NPN Darlington transistor such as the BC517.
- If you place two wires in a glass cup, you can monitor any kind of solution. Adjust P1 to have a half-scale deflection. You then can detect changes in the resistance due to paranormal phenomena. Use a solution with biological components (bacteria or lacto bacillus, for instance) and see how it can react in ESP or telekinesis experiments.

Parts List: Project 28

Semiconductors

Q1, Q2   BC548 or equivalent general-purpose NPN silicon transistor

Resistors

R1   100 kΩ, 1/8 W, 5%—brown, black, yellow

R2   4.7 kΩ, 1/8 W, 5%—yellow, violet, red

R3, R4   1 kΩ, 1/8 W, 5%—brown, black, red (4.7 kΩ—yellow, violet, red—for 9V supply)

Miscellaneous

P1   10 kΩ potentiometer (47 kΩ for a 9V battery)

M1   50 to 200 μA current meter, zero in the center of the scale (see text)

B1   3 to 9 V, AA cells or battery

X1, X2   electrodes (see text)

Terminal strip, cell holder or battery clip, box, knob, wires, solder

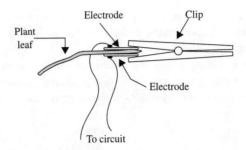

Figure 142 Experimenting with plants.

Project 29: Hypnotic LEDs

This circuit can be used for meditation training or to control your paranormal abilities, and it is applicable to many paranormal experiments as suggested in the text. The hypnotic LED circuit is formed by a low-frequency oscillator with its frequency controlled by the resistance between two electrodes. The LEDs can be mounted on eyeglasses or as needed for a particular experiment.

From changes in skin resistance or in the resistance of a sensor attached to any living thing (e.g., a leaf), the circuit alters the flash rate of two LEDs.

To fabricate a hypnotic biofeedback device, the LEDs can be mounted on eyeglasses as shown in Fig. 143. Many other paranormal experiments can use this circuit as suggested in the next section.

Experiments

- The hypnotic LEDs can be used in experiments to see how the stress produced by LED flashes can alter the ESP abilities of a subject or induce him into a

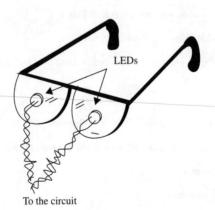

Figure 143 Hypnotic glasses using two LEDs.

trance state or fourth state of consciousness. It also can be used as a monitor when the LEDs are used by the researcher and the subject acts on the sensor.
- The flashes produced by the LEDs can be used to monitor the influence of the subject on the circuit. Changes in the flash rate must be produced by the subject using mental power only. Sensitive objects and living things can be placed between the electrodes as in other experiments described in this book.
- Variations in the resistance of a leaf or the earth in a flower pot can be monitored using the flash rate of the LEDs. In many experiments, the flash rate can be used as some form of feedback. For instance, you can place the plant that controls the flash rate in a lightproof box that is illuminated by the LEDs to create a circadian (i.e., 24-hour cyclical) rhythm experiment.
- The influence of a pendulum on a living being can be monitored via the flash rate variations.

How It Works

The circuit is formed by two low-frequency oscillators made with the NAND Schmitt trigger gates of a 4093 CMOS IC. The basic frequency of each circuit depends on the capacitor in the input (C1 and C2) and on the feedback network.

In the first oscillator, the feedback network is formed by R1 and C1, and its frequency can be adjusted by P1. In the second oscillator, the feedback network is the resistance between the electrodes.

The signals generated by the two oscillators are combined in two LEDs such that LED1 glows when the IC's pin 10 is high and pin 4 is low. If pin 10 is low and pin 4 is high, LED2 will glow.

As the logic levels in the outputs (pin 4 and 10) change continuously, depending on the frequency and phase of the generated signals, the LEDs will flash alternatively at a random rate.

The aim of the circuit is to control the random rate, thereby imparting to the flashes a definite pattern. This is accomplished by changing the feedback of the second oscillator, where the sensor is installed.

The fast flash rate of the LEDs can be used both to monitor and control the physiological functions of the body in a feedback application, but they have other applications. Many variations can be made to the original circuit as described in the suggestions section.

Assembly

Figure 144 shows the complete schematic diagram of the hypnotic LEDs. The circuit is mounted on a small printed circuit board as shown in Fig. 145. R2 depends on the power supply voltage as shown in the following table.

Power Supply Voltage	R2
3 V	220 Ω
6 V	470 Ω
9 V	1 kΩ

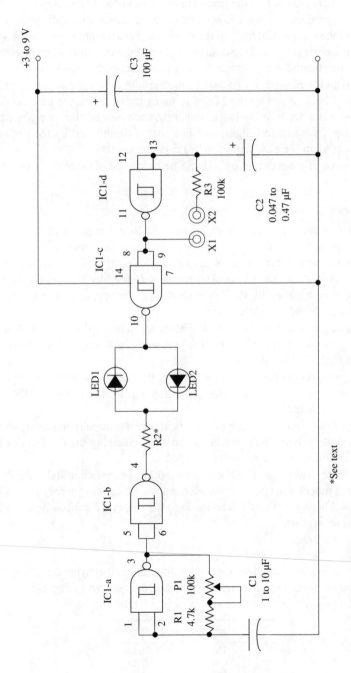

Figure 144 Hypnotic LEDs.

Experimenting with Paranormal Skills 193

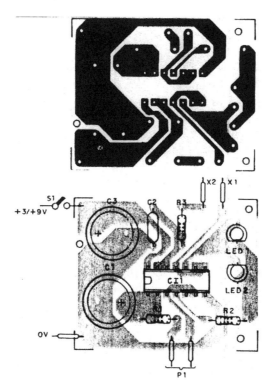

Figure 145 Printed circuit board for Project 29.

For better results, make LED1 red and LED2 green. Capacitor C1 values can be chosen by the reader to suit the experiment. Capacitors between 1 and 10 µF can be used in this project. The value of C2 is also set experimentally to give a 50% duty cycle for the LED flashes when the subject's fingers are placed on the electrodes or when the subject is plugged into X1 and X2. Electrodes intended for biofeedback applications can be made from two metal rods.

The circuit can be housed in a small plastic or wooden box. Dimensions are basically determined by the power supply.

Using the Circuit

We can test the circuit in a basic biofeedback application by plugging two metal rods into the input as electrodes. Holding one in each hand, turn on the power supply. Then adjust P1 to have a flash rate in which the on time of LED1 is approximately equal to the on time of LED2 (which varies according to applied pressure).

Now try to keep the flash rate constant. Do not alter the pressure of your hands on the electrodes—do it only by concentrating on some vital function such as your blood pressure or skin resistance. Try to control that function by willpower.

The simplest application of the device is as a feedback circuit to train people to control their physiological functions. But many other experiments can be performed using this circuit.

One interesting configuration is a "plasma" sensor made from a candle, as shown in Fig. 146. The flame is an electric current conductor, and it forms the feedback loop that keeps the circuit in oscillation. In the case of a candle, the mystical aspect can contribute to some interesting experiments. The resistance of the candle varies constantly, altering the frequency of the oscillator. Try to fix your mind on the flame and alter the frequency of the circuit or control the flashing of the LEDs.

Suggestions

- Alter the value of C1 and/or C2 to change the flash rate of the LEDs. Lower values can give you a stroboscopic effect (higher flash rate) that can be used in other experiments.
- Place two other electrodes in series with P1 to produce "differential biofeedback." You can control the flash rate by concentrating on both oscillators.
- Place two electrodes in a conductive solution and monitor how the flash rate can be controlled by a subject's mind in a PK experiment.
- Replace the LEDs with infrared types and perform experiments in ESP with "sensitive" subjects to determine if they can sense the infrared emissions.

Parts List: Project 29

Semiconductors

IC–1	4093 integrated circuit
LED1, LED2	Common red/green LEDs (see text)

Resistors

R1	4.7 kΩ, 1/8 W, 5%—yellow, violet, red
R2	220 Ω to 1 kΩ according the power supply voltage (see text and table)

Capacitors

C1	1 to 10 µF/12 WVDC, electrolytic (see text)
C2	0.047 to 0.47 µF, metal film or ceramic (see text)
C3	100 µF/12 WVDC, electrolytic

Miscellaneous

P1	100 kΩ potentiometer
X1, X2	electrodes (see text)

B1	3 to 9 V, two or four AA cells or a 9 V battery (see text)
S1	SPST, toggle or slide switch

Printed circuit board, plastic box, cell holder or battery clip, wires, solder, etc.

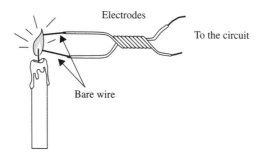

Figure 146 Controlling the circuit by concentrating on a flame.

Project 30: Electro-Shock Generator

From the information picked by the sensors, this circuit produces an electric shock that changes in frequency and also in voltage. Because the circuit can be adjusted to any safe voltage value, it is not dangerous.

The central frequency (for pulses or a tone) and the intensity can be adjusted so as to make the device appropriate for many paranormal experiments. This means that the voltage output can be placed between values from the lower limit of feeling up to an unpleasant shock.

The experiments available to the reader using this device are basically the same as described in previous projects. The difference is that the output is used to stimulate the skin of a subject. Some suggested experiments are described below.

The circuit is powered from common cells.

Experiments

- The presence of an electrical stimulus in the subject, based on changes in his own skin resistance, can be studied in ESP experiments. The reader can study how this "external" input alters (increases or decreases) the paranormal abilities of an individual.
- Sensitive objects or living beings can be placed between electrodes, and the subject will be *auto-stimulated* by the voltage produced by the circuit. Experiments to control this voltage using mental powers can be made using this circuit. The circuit can be used to produce current fields in flower pots or biological conductions produced by their own stimulus.

- The circuit can be used to help the subject to reach the *fourth state of consciousness* or to go into a trance state. The aim of the experiment is to keep the output voltage and frequency constant without altering the pressure of the hands or fingers on the electrodes. Many experiments in transcendental meditation and biofeedback can be conducted based on this configuration.
- Place sensitive objects on the electrodes and try to alter the output voltage by focusing your mental powers on the subject. Experiments in telekinesis can be made by adding a few variations to this basic configuration.
- In all the experiments suggested above, the influence of a pendulum can be explored when working with radiesthesia.
- The output of the circuit can be applied to the leaf of a plant. The electrodes will be placed on other leaf. The experimenter can study how the plant itself might control the voltage of the circuit.
- Plugging the circuit into a fluorescent lamp, experiments with modulated light are possible.

How It Works

A gate of a 4093 IC is used as an oscillator where the frequency is determined by C1 or C2, R1, and the resistance between the terminals X4 and X5, where the electrodes are placed. As the resistance between X4 and X5 changes, the frequency also changes.

The square signal produced by this block is applied to the input of three other NAND gates that exist in the 4093 and are wired as buffer-inverters. The signal at the output of this digital amplifier is then applied to a power NPN Darlington transistor.

The load of this transistor is the low-voltage winding of a transformer. The pulses applied to this winding appear as high-voltage pulses in the other winding and are applied to terminals C1, X2, and X3 where electrodes are connected. One of the electrodes is placed in X2, where the potentiometer P1 can be used to control the intensity of the stimulus.

The circuit is powered from AA cells but, using common power supply transformers, high-voltage pulses up to 300 V can be produced. Of course, these pulses are not dangerous, as they are short and limited in current. Otherwise, the pulses are strong enough to drive a fluorescent lamp that will flash according the resistance between electrodes. Many experiments can be performed using this circuit.

Assembly

Figure 147 shows the diagram of the high-voltage generator used for electroshock. The components can be placed on a small printed circuit board as shown by Fig. 148.

For Q1, you can use any medium-power Darlington transistor with collector current rated to 1 A or more. This transistor must be installed on a small heatsink.

Experimenting with Paranormal Skills 197

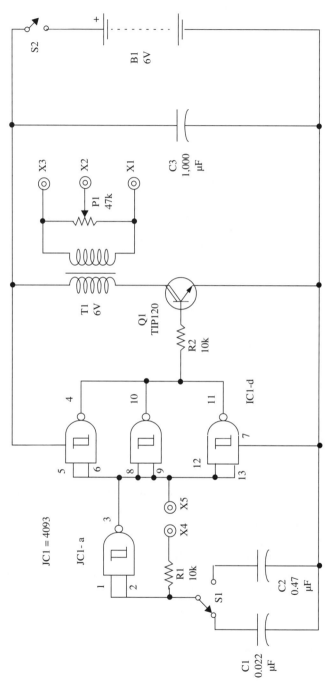

Figure 147 Electro-shock generator.

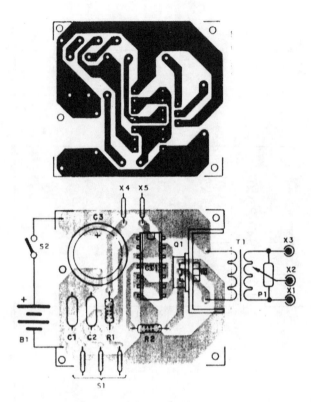

Figure 148 Printed circuit board for Project 30.

Any transformer with a secondary rated to voltages between 5 and 12 V and currents in the range between 50 mA and 300 mA can be used. The primary can be rated for 117 Vac or 220/240 Vac.

The type of electrodes required depends on the experiment.

As a sensor, you can use small metal plates upon which the fingers are placed, or metal rods. For the stimulation electrodes, you can use small metal plates fixed to the skin by rubber bands or even rods to be grasped in the subject's hands.

In some cases, you need to add a 2.2 MΩ resistor between X4 and X5 to keep the circuit from going into a non-oscillating condition. This condition can increase the current produced by Q1, producing excess heat and draining the cells.

The circuit can be placed in a small plastic or wooden box and to the connection of the electrodes bornes or banana jacks can be used.

Testing and Using the Circuit

Insert a 100 kΩ to 1 MΩ resistor between X4 and X5 and connect a pair of metal plates as electrodes between X1 and X2. Set the potentiometer P1 to its minimum output. S1 must be placed in a position to connect C1 to the circuit. Turn on the circuit and place your fingers on the electrodes.

Open P1 slowly until you feel the signal produced by the circuit, first as a slight sensation of itch. Continue until the sensation becomes a light shock. (Don't continue to the point where it becomes painful.) Now, move S1 to the position in which C2 is connected to the circuit and experiment it again.

The next step is to replace the resistor between X4 and X5 with a sensor and perform the experiments.

When in operation, let the subject adjust P1 by himself. Don't forget to adjust it to the minimum value before starting with the probes!

Suggestions

- Place an LDR between X4 and X5 to make experiments with light (see the next project, "Third Eye").
- A fluorescent lamp can be connected between X3 and X4 to perform experiments with a visual feedback.
- C2 can be increased to values as high as 4.7 μF to allow the circuit to produce widely spaced pulses. In this case, add a 47 Ω resistor in series with the transistor emitter to avoid draining excess current from the cells.
- Transistor Q1 can be replaced by any power FET without making changes in the original circuit.
- Replace T1 with a coil formed by 20 to 100 turns of AWG 28 wire in a cardboard form as shown in Fig. 149. The circuit can be used to apply magnetic fields to subjects as controlled by the resistance between X4 and X5.
- The fluorescent lamp placed between X1 and X3 can be of the "black light" type, such as the ones used in night clubs. Experiments with UV can be performed this way.

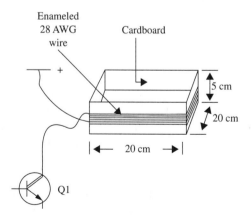

Figure 149 Generating a magnetic field with a cardboard coil.

Caution: Do not use UV lamps of the types recommended for erasing EPROMs. They are dangerous!

Parts List: Project 30

Semiconductors

IC1	4093 CMOS integrated circuit
Q1	TIP120, medium-power silicon NPN transistor

Resistors

R1, R2	10 kΩ, 1/8 W, 5%—brown, black, orange

Capacitors

C1	0.022 µF, ceramic or metal film
C2	0.47 µF, ceramic or metal film
C3	1,000 µF/12 WVDC, electrolytic

Miscellaneous

P1	47 kΩ potentiometer
T1	Transformer: primary 117 Vac, secondary 6 V, CT, 250 to 500 mA
S1	SPST, toggle or slide switch
S2	One pole, two positions switch
B1	6 V—four AA or C cells
X1, X2, X3	Output electrodes (see text)
X4, X5	Input electrodes (see text)

Printed circuit board, plastic box, cell holder, electrodes, knob for P1, wires, solder, etc.

Project 31: Third Eye

Experiments with paranormal senses, such as in ESP, invoke the existence of what is called a *third eye*. This eye is analogous to the physical eyes that look out from the front of our heads, except that it provides the ability to "see the invisible" and to discern things that are displaced in time and space.

This project is a simulation of that eye, using a light-dependent resistor (LDR) as a sensor. It is fundamentally derived from the last project.

Basically, the project uses the LDR to "sense" light or images in an ambient and produce from them an electric signal that can be used to excite subjects as shown in Fig. 150.

Experimenting with Paranormal Skills 201

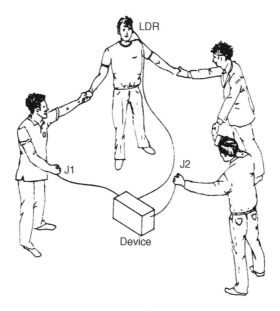

Figure 150 A paranormal experiment with the "third eye" involving four people.

The excitations consist of high-voltage signals (pulses or tones) applied to electrodes that can be touched by the subject or subjects. Figure 150 shows an experiment in which the subjects form a circle and, by joining hands, close the circuit such that all receive the stimulus produced by the circuit.

We should observe that the reader might have the false impression that the person who is touching the electrodes receives the most powerful discharge. This is wrong. As the current flowing in the circuit is the same throughout, all persons receive the same discharge.

The circuit is powered from AA cells or a 9 V battery and can be placed in a very small plastic or wooden box for portable use. Current drain is very low, allowing the life of the battery or cells to extend many weeks or even months.

Experiments

- Experiments in ESP involve stimulation of the subject using the third eye. Using the device with several people as suggested in Fig. 150 can result in an interesting experiment.
- As in other experiments, the subject can try to alter the pulse rate or output voltage using his mind as a PK experiment. Light sources can be used as target of the subject with PK powers, and the LDR is used to detect changes in light intensity. In particular, we recommend using a candle because of its mystical implications.

- For transcendental meditation and biofeedback, this device can help the subject find the *fourth state of consciousness* or trance. The LDR can be placed in other locations such as focused on a point on the wall or a distant light source.
- In radiesthesia experiments, the influence of the pendulum on the subject or on an LDR placed in any location can be studied.
- The LDR can be excited by the flickering flame of a candle to conduct an interesting experiment involving paranormal forces, ghosts, and spirits.

How It Works

The circuit uses the same oscillator stage as the previous projects. One NAND gate of the four existing in a 4093 IC is used to produce pulses at a rate determined both by C1 or C2 and the resistance of an LDR or CDs cell.

The amount of light falling on the LDR determines its resistance and therefore the frequency of the circuit. Frequency rises along with light intensity.

With C1 in the circuit, we have frequencies in the range of 200 and 5000 Hz. Installing C2 in the circuit, the frequency becomes low—in the range of 0.2 to 5 Hz, producing interval pulses. The signals produced by this stage are applied to the three other gates that are wired as a digital amplifier (buffer-inverters).

The output of the amplifier stage drives a PNP medium-power transistor. As the load, this transistor has the low-voltage winding of a transformer. In the high-voltage secondary, we have the signals that are used to stimulate the subjects.

Using a common 117 Vac transformer, the output voltage can reach values up to 300 V. But, in this case, the current is very low, avoiding any danger to the subjects—although the sensation of shock can rise to very uncomfortable levels. That is why we added a voltage control, formed by P1 in the output.

Assembly

The complete diagram of the third eye is shown in Fig. 151. The circuit can be mounted on a small printed circuit board and housed into a plastic or wooden box as shown by Fig. 152.

Any medium-power PNP transistor with collector current rated to 1 A or higher can be used in this project. The transistor must be mounted on a small heatsink. This heatsink is formed by a piece of metal bent to form a "U" or an "L."

The LDR is another non-critical component. Any type with any diameter can be used in the project. The LDR can be mounted on a support to be affixed to the subject's head and installed in a small cardboard tube. If a convergent lens is placed in front of the LDR, we add directivity and sensitivity to the circuit. It will be able to pick up a small amount of light coming from distant or weak light sources.

The electrodes are made as in the previous project. You can use small metal plates affixed to the skin with a rubber band or, if you prefer, two metal rods placed in your hands.

The box dimensions depend on the power supply, as a battery is smaller than a holder for four AA cells.

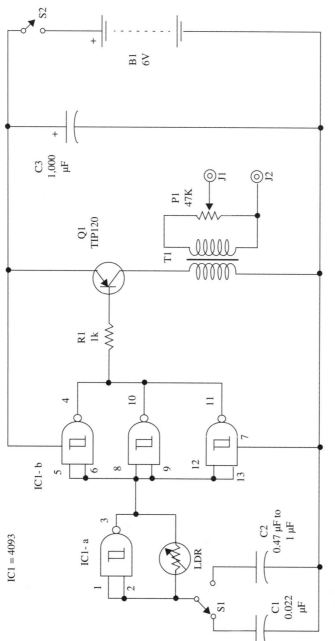

Figure 151 Third eye.

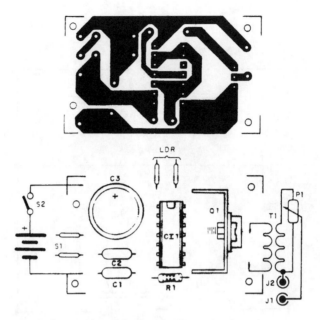

Figure 152 Printed circuit board for Project 31.

Testing and Using the Circuit

Place the electrodes on any part of your body or, if they are rods, hold them in one hand, but do not let one touch the other. Adjust P1 to the lowest output voltage. Place the LDR near any light source and turn on the power supply. Place S1 in the position that connects C1 to the circuit. Test the circuit again with S1 in the other position.

Open P1 until you begin to feel a slight itch. Pass your hand in front of the LDR to see if the circuit changes its performance when the amount of light reaching the sensor is reduced. Then, to test the output of the circuit, open P1 more until the shock comes uncomfortable.

Now you can use the circuit in the experiments. Remember: every time you use the device, you must start the experiments with P1 closed. If you don't, a potent shock will be applied to the subject when the circuit is turned on.

Note that you can also test the circuit using a neon lamp as indicator. When placed at the output of the circuit, it will glow when the voltage peaks at 80 V. A fluorescent lamp will also glow if plugged into the output of this circuit.

Suggestions

- You can use only one value for C1. This way, C2 and S2 are not necessary in this project.

- Powering the circuit from a 9 V or 12 V supply, you can increase the output power to use the circuit with fluorescent lamps. **Do not use the circuit with human subjects if powering it from ac power supplies fed from an ac power line.**
- Replace the transformer with a coil formed by 20 to 100 turns of 28 AWG wire on a cardboard form. The circuit can be used to produce a magnetic field from the light on the sensor. This magnetic field can be applied to subjects of PK or ESP experiments.
- Replace the transistor with any power FET such as the IRF640. No changes to the circuit are necessary to get the acceptable performance.

Parts List: Project 31

Semiconductors

IC1	4093 CMOS integrated circuit
Q1	TIP32 medium-power PNP transistor

Resistors

R1	1 kΩ, 1/8 W, 5%—brown, black, red

Capacitors

C1	0.022 µF, ceramic or metal film
C2	0.47 µF to 1 µF, ceramic or metal film
C3	1,000 µF/12 WVDC, electrolytic

Miscellaneous

LDR	Any light-dependent resistor or CdS cell (see text)
P1	47,000 Vac × 6 to 12 V × 50 to 300 mA transformer (see text)
S1	One pole × two positions, toggle or slide switch
S2	SPST, toggle or slide switch
J1, J2	Banana jacks
B1	6 V, four AA cells

Printed circuit board, plastic or wooden box, knob for P1, electrodes, cell holder, wires, solder, etc.

Project 32: Hypnotic Glasses

In this project, we create a device intended for several paranormal experiments. It consists of a common pair of glasses in which the lenses are replaced by two bi-color, flickering LEDs as shown in Fig. 153.

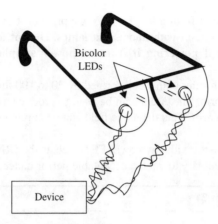

Figure 153 Bicolor LEDs are used in this experiment.

Placed near the subject's eyes, the flickering effect of the LEDs can induce him into trance, fourth state of consciousness, or simply into a relaxed state. The LED flashes are powerful enough to be sensed even when the subject closes his eyes.

In addition to applications aimed at relaxation or hypnosis, the circuit can be adapted for use in other paranormal experiments. For instance, the visible light LEDs can be replaced by IR (infrared) devices to see if they have any effect on paranormal experiments involving ESP. They can be used to determine if the subject can sense radiation at wavelengths outside of the visible spectrum.

The circuit is powered from AA cells or a 9 V battery and is housed in a small plastic box, so it can be transported to any desired location for the experiments.

With appropriate modifications, the following experiments can be performed with this project.

Experiments

The hypnotic glasses can be used to assist the subject in achieving a special mental state in experiments that aim for relaxation, the fourth consciousness state, or even a trance. An interesting suggestion is the use of infrared LEDs to see if invisible light can have some effect on the subject or if he can detect it via a "sixth sense." The LEDs can also be mounted on a panel. The subject must concentrate on them to achieve the special mental state required for the experiments.

Although the frequency or flash rate of the LEDs is an intrinsic property of the circuit, in a PK experiment, the subject can try to alter it with his mind. Another person involved in the experiment (a sensitive) can try to detect those changes by observing the flashing LEDs. The circuit can also be used to help the subject to achieve a required mind state before starting with other experiments.

In radiesthesia experiments, the effects of flashing LEDs on a person can be detected or monitored via the use of a pendulum. The researcher can also try to use a pendulum to change the flash rate.

In transcendental meditation and biofeedback, the fourth state of consciousness can be achieved using this circuit as described in the beginning of the text as a basic application for this circuit. The experimenter can also investigate how the LED flashes can affect the time required to enter the special mental state.

In experiments involving plants and animals, the flashing LEDs can be used as a stimulus to produce some paranormal effect. For example, how will a plant leaf react to some kind of paranormal stimulus if it is placed in a dark box and illuminated only by the flashing LEDs?

How It Works

This circuit is also based on an oscillator using the 4093 CMOS IC. One of the four gates of the IC is used as a low-frequency oscillator. The frequency range of about 0.1 to 5 Hz is determined by C1 and adjusted by P1. The digital signal produced by this stage is applied to two other stages using the remaining gates of the IC.

One stage is formed by a single gate that operates as a buffer-inverter, driving the LED of one color (green, for instance). The LED will glow when the output of this stage goes to the high logic level.

The other stage is formed by two NAND gates of the 4093 IC that are also wired as buffer-inverters. This way, as the applied signal is inverted and uninverted again, the output of this stage goes to the high level when the output of the other stage goes to low. This stage drives the other color LED (red, for instance)

This means that when one output is high, the other is low, and vice versa. As the circuit changes its state at a rate determined by the oscillator, the LEDs flash in an alternating manner.

Assembly

Figure 154 shows the complete schematic diagram of this device. The project can be mounted on a small printed circuit board as shown by Fig. 155.

Bicolor LEDs (red and green or red and yellow) are suggested, but if you have difficulty in finding them, you can use common LEDs. Any color can be used to suit the experiments you have in mind. You can even test different combinations of colors in your experiments.

R4 and R5 determine the brightness of the LEDs. For a 6 V battery, we recommend a 470 Ω resistor. For a 9 V battery, we recommend a 1 kΩ resistor. But, in some experiments, the amount of light produced by the LEDs must be reduced. Very low light levels are necessary to avoid sensory overload to the subject's eyes. To accomplish this, you only have to increase R4 and R5 to values up to 22 kΩ. Higher values also imply lower current drain from the power supply, extending its life.

The circuit can be installed in a small plastic box with the LEDs placed on common glasses, attached to the lenses. Wires 100 cm in length are used to connect the circuit to the glasses.

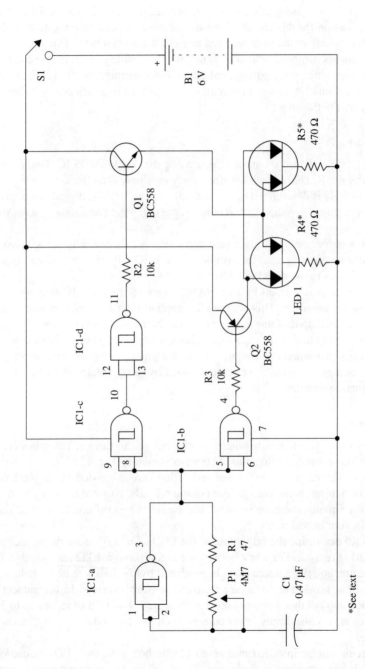

Figure 154 Hypnotic glasses.

Experimenting with Paranormal Skills

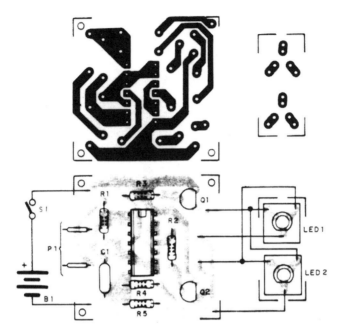

Figure 155 Printed circuit board used in Project 32.

Testing and Using the Circuit

Turn S1 on and see if the LEDs flash. Adjust the frequency using P1. If the desired flash rate is not found when adjusting P1, replace C1. Values between 0.047 and 1 µF can be tried.

When using the circuit, adjust P1 to the desired flash rate to suit the experiment. Alter the brightness of the LEDs if necessary.

Suggestions

- Replace P1 with a touch sensor and make a biofeedback loop that can be used in many other paranormal experiments.
- P1 can also be replaced with an LDR. The flash rate then will depend on the amount of light striking the LDR.
- A seven-segment digital display can replace the LEDs as shown in Fig. 156, producing programmed, alternating symbols. You can program the circuit to produce "0" and "1" alternately and use the circuit in ESP experiments.
- Replace the transistors with power FETs and power the circuit from 12 V supplies. The LEDs and the series resistors can be replaced with incandescent 12 V lamps up to 500 mA.
- The LEDs can be replaced with a small incandescent lamp rated for 6 or 9 V × 20 to 50 mA. The current-limiting resistor is not necessary in this case.

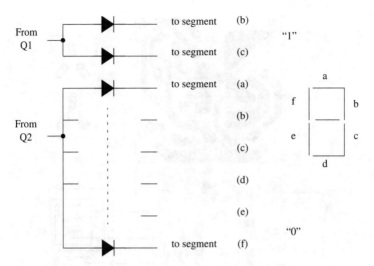

Figure 156 Displaying "0" and "1".

Parts List: Project 32

Semiconductors

IC1	4093 CMOS integrated circuit
Q1, Q2	BC558 general-purpose PNP transistors
LED1, LED2	Two bicolor LEDs or four common LEDs (see text)

Resistors

R1	47 kΩ, 1/8 W, 5%—yellow, violet, orange
R2, R3	10 kΩ, 1/8 W, 5%—brown, black, orange
R4, R5	470 Ω, 1/8 W, 5%—yellow, violet, brown, according to the power supply voltage (see text)

Capacitors

C1	0.47 µF ceramic or metal film (see text)

Miscellaneous

P1	4.7 MΩ potentiometer
S1	SPST, toggle or slide switch
B1	6 or 9 V, four AA cells or battery

Printed circuit board, plastic or wooden box, knob for P1, cell holder or battery clip, wires, solder, etc.

Project 33: Bioprobe

This is a portable *bioamplifier*, a small circuit that can be used to detect biological signals and convert them into an audible tone. Changes in skin resistance, data from a sensor, or signals picked up from plants can be converted into an audible sound for reproduction by a loudspeaker. Paranormal phenomena can be investigated by whatever electric signal they produce via an appropriate sensor.

The circuit is very small and highly portable. It is powered by common AA cells and has a very low current drain.

Several applications in experiments involving paranormal phenomena are suggested, starting with the use of the device as described below.

Experiments

The electrodes or sensors plugged into the input of this circuit are determined by the kind of experiment you want to perform. You can connect the circuit directly to the skin of a subject to detect any kind of electric signal generated by his body in ESP, PK, or transcendental meditation experiments. Two metal plates or rods can be used as electrodes. We remember that the nervous cells of a subject produce voltage signals in the range of millivolts. It is these voltages that are detected by electrocardiographs and electroencephalographs. You can also plug a coil into the circuit to pick up magnetic fields, converting signals into sounds for ghost detection and other paranormal experiments.

Plug a coil (1,000 to 5,000 turns of 28 to 32 AWG wire in a ferrite core) into the input and try to use your mind or the mind of a subject to induce signals. The circuit can also be used to detect signals in the body of a subject during experiments involving PK.

In radiesthesia experiments, the influence of the pendulum on the body of a subject can be detected from the alterations in signals picked up with this amplifier. The experimenter can also plug a pickup coil into the input, as suggested for PK experiments, to see the influence of a pendulum.

In transcendental meditation and biofeedback, the subject can use signals from his own body and convert them into sounds to control internal physiological functions, as in other experiments involving temperature-to-sound or temperature-to-light converters.

In summary, whatever the reader wants to plug into the input of this device is a matter of choice. Phototransistors, pickup coils, microphones, and many other transducers and sensors can be used in these experiments.

The circuit can also be used as a remote-listening aid. By placing a microphone in another room (use a shielded cable), the experimenter can hear sounds produced by paranormal events. Use a high-impedance microphone to stimulate this amplifier.

Two metal rods or pieces of bare wire can be used to pick up sounds from the earth or water.

How It Works

The circuit has a very simple configuration based on a common bipolar transistor. Three stages are used to amplify the low-frequency signal that is applied to the input. The first transistor is wired in a common collector stage, and the next two form a two-stage Darlington-coupled amplifier.

The very simple configuration has the advantage of being very easy to mount, even by beginners or persons less experienced with electronic mounting techniques.

Assembly

Figure 157 shows the diagram of the bioprobe. The circuit is mounted using a small printed circuit board as shown by Fig. 158.

Observe the position of polarized components such as the transistor. For the input, you can use a jack or, if you prefer, two pieces of wire with alligator clips on the end. For the connection of the earphones or headphones, use appropriate jacks. The dimensions of the box are basically determined by the size of the printed circuit board and the cell holder.

Testing and Using the Circuit

Into the circuit input, plug a high-impedance microphone or a coil formed by 1,000 to 10,000 turns of any wire on a ferrite rod. Plug an earphone or a headphone into the output.

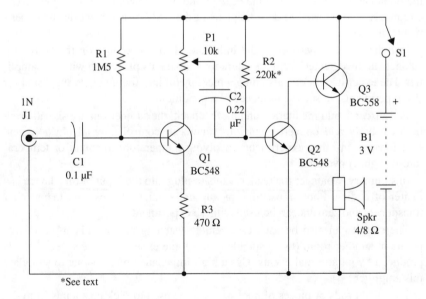

Figure 157 Bioprobe.

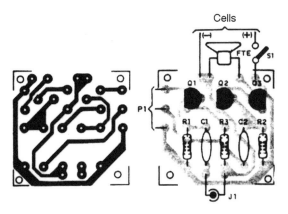

Figure 158 Printed circuit board for Project 33.

Turn on S1 and open P1 until you can hear something. In the case of the microphone, you'll hear the ambient sound. In the case of the coil, you will pick up the "hum" of the ac power line or, in some cases, signals of some nearby radio stations (they will be mixed, as the circuit is not tuned).

If the sound is distorted when you open the volume control (P1), you must alter the values of R1 and R2. R1 can be varied in the range between 1.5 and 2.2 MΩ. In some cases, a resistor of even less than 1.5 MΩ must be used. R2 should be in the range between 220 and 470 kΩ.

When using the circuit, you only have to plug the sensor or transducer into the input of the circuit. If the signals are too low, it is because the transducer or sensor does not produce enough power to drive the circuit.

Suggestions

- Replace Q3 with a TIP32 or BD136 and power the circuit from a 6 V or 9 V power supply to increase the output power. In this case, you can use a loudspeaker as the transducer.
- Add a 0.01 to 0.047 µF capacitor in parallel with the input to reduce the high-frequency (treble) sounds.
- Using the pickup coil, you can detect magnetic fields produced by the ac power line.

Parts List: Project 33

Semiconductors

Q1, Q2   BC548 or equivalent—any general-purpose NPN silicon transistor

Q3   BC558 or equivalent—any general-purpose PNP silicon transistor

Resistors

R1   1.5 to 2.2 MΩ, 1/8 W, 5% (see text)

R2   220 kΩ to 470 kΩ, 1/8 W, 5% (see text)

R3   470 Ω, 1/8 W, 5%—yellow, violet, brown

Capacitors

C1   0.1 μF, ceramic or metal film

C2   0.22 μF, ceramic or metal film

Miscellaneous

P1   10 kΩ, potentiometer

S1   SPST, toggle or slide switch (or ganged to P1)

B1   3 V, two AA cells

HP   Low-impedance earphone or headphone

J1   Input jack or two alligator clips

Printed circuit board, battery holder, plastic box, knob for P1, wires, solder, etc.

Project 34: Paranormal Electroscope

The presence of electric charges in an ambient, a subject, or an object during a paranormal experiment can reveal important facts that are useful to the researcher. Can a subject alter the electric charge of a part of his body during a PK experiment? Is the electric charge of the ambient around the subject affected when he achieves the fourth state of consciousness or trance?

These questions can be answered if the experimenter has a device that can detect electric charges, i.e., an electroscope. The electroscope described here is extremely sensitive and, at distances of up to a meter, can detect the static electric charges produced when, e.g., you rub a pen against your clothes. The electroscope can be used in several paranormal experiments as suggested below.

Experiments

Placing the electroscope near the subject of a PK experiment, it is possible to detect any change in the ambient electric charge. Another interesting experiment can be made by connecting the electroscope to an object. The researcher can detect, via the electroscope, if the electric charges in this object can be changed by willpower. The subject will try to alter the indication of the electroscope using his mental powers. The object can be wired to the input of the circuit, and the subject must alter the properties of the object or move it across the surface of a table. The researcher can monitor any changes that appear in the electric charges during the PK experiment.

In any transcendental meditation and biofeedback experiment, monitoring the electric charge in the subject's body or in the ambient is important. The subject can try meditating while seated on an isolated platform as shown in Fig. 159. The electroscope will indicate any change in his electric charge during the meditation state or trance.

The influence of a pendulum on any object or person can affect not only physiological states but also physical quantities not seen or felt by the researcher. This includes the magnetic and electric fields produced by accumulated charges. Changes in the electric charge can be detected by placing this device near the object or person or by connecting it to the object or subject. In the case of a human, the person can be placed in an isolated platform as suggested in Fig. 159.

UFO researchers say that the presence of a flying saucer or any strange ship can alter magnetic and electric fields around them, causing interference in many appliances. The presence of a UFO that is too distant to be detected by the malfunction of appliances still can be detected by an electroscope or a magnetic field detector. (The latter is described in the section on "UFOs and Ghosts.") The circuit can be wired to a small antenna, as shown in Fig. 160, to detect changes in the atmospheric charge. This circuit will also detect the presence of charged clouds passing over your home when a storm is forming.

Warning: Do not leave the circuit connected to the antenna during a storm, as it can act as a lightning rod!

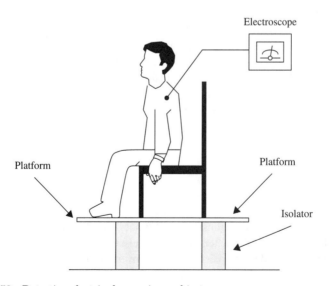

Figure 159 Detecting electric charges in a subject.

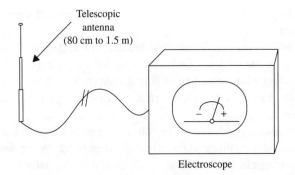

Figure 160 Detecting charges in the air and clouds.

In experiments involving the presence of ghosts or poltergeists, the electroscope is a powerful tool for detecting their presence.

How It Works

The current flowing between the source (s) and the drain (d) in a junction field-effect transistor (JFET) depends on the voltage applied to the gate (g). As the gate is isolated from the channel in which current flows, the device has a very high input impedance. This means that small electric charges in the gate produce electric fields that can control the current flow.

If an electrode is wired to the input of the circuit, any electric charge in this electrode can induce charges in the gate region and alter the current flow through the transistor. In the electroscope, the electrode is the sensor, and the transistor is wired to form a branch of a Wheatstone bridge.

Any change in the charges detected by contact or induction can alter the bridge equilibrium, as will be indicated by a current meter. The user has only to put the bridge in equilibrium and observe the meter indications.

A problem to be avoided is that high voltage applied to the gate of the transistor can burn it. The spark between the gate and the channel region produced when high voltage is applied destroys the device. So **never** touch any charged object with the electrode!

Assembly

Figure 161 shows the diagram of a simple paranormal electroscope. Any common JFET general-purpose transistor will function in this circuit. We recommend the BF245 as shown in the schematic. If you use the MPF102, the terminal placement is different, and the transistor must be inverted (flat part down).

The ideal meter for this project is a 50 μA unit with zero in the center of the scale. This type of meter is used in some devices as a battery state indicator or

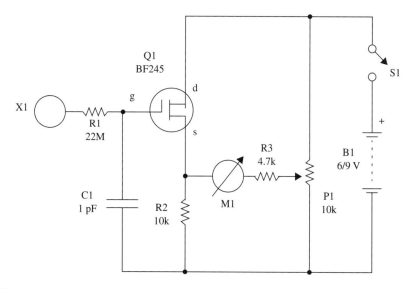

Figure 161 Paranormal electroscope.

VU meter. But if you can't find one, a common meter with full scale values between 50 and 200 µA can be used. You will just have to adjust the equilibrium point to a current other than zero when using the circuit.

The electrode is a loop of bare wire or a metal sphere. The circuit can be housed in a small box (plastic or wood) and used anywhere.

Testing and Using the Circuit

Install the cells or battery and turn on S1. Adjust P1 until the needle of the meter reaches approximately the middle of the scale. Then rub a pen (plastic, spherical) over the surface of your clothes and place it near the electrode without touching it. The needle will move, indicating the presence of electric charges. *Note: remember that, on humid days, it is difficult or impossible to charge objects.*

When using the circuit, you only have to adjust P1 to set the needle to the middle of the scale and observe any movement indicating the presence of charges.

Suggestions

Plug an external antenna into the circuit to detect charges in the air or in the clouds. The wire must be short in this application. If you can't find a meter suitable for this application, use your multimeter set to the lowest current scale.

The device can be installed inside a glass bottle, imitating the classic gold metal plate electroscope found in physics laboratories.

Parts List: Project 34

Semiconductors

Q1   BF245 or MPF102 JFET (see text)

Resistors

R1   22 MΩ, 1/8 W, 5%—red, red, blue

R2   10 kΩ, 1/8 W, 5%—brown, black, orange

R3   4.7 kΩ, 1/8 W, 5%—yellow, violet, red

Capacitors

C1   1 pF, ceramic

C2   0.1 µF, ceramic or metal film

Miscellaneous

P1   10 kΩ potentiometer

X1   Electrode (see text)

M1   50-050 µA microampere current meter (see text)

S1   SPST, toggle or slide switch

B1   6 or 9 V, four AA cells or a 9 V battery

Terminal strip, plastic or wooden box, knob for P1, cell holder or battery clip, wires, solder, etc.

3.6 Paranormal Experiments with Light

> *Light seeking light doth light of light beguile.*
>
> William Shakespeare (1564–1616)

Light is a form of electromagnetic radiation that can be used to perform some interesting experiments in paranormal sciences. Monochromatic light sources, white light sources, modulated light sources, infrared or ultraviolet sources, polarized light sources, coherent light sources (lasers), and many others can be used to detect or to cause paranormal phenomenon, as suggested in the next projects.

The special light sources described here can be used with many of the other projects, adding new effects and helping the researcher to make new discoveries.

Project 35: Simple Light Detector

Some paranormal phenomena can involve light variations too slight to be seen by the human eye. A device with greater sensitivity than the human eye, and one that also can detect UV and IR light, is a very useful tool in the paranormal experimenter's lab. The circuit described here is a very sensitive photodetector powered from AA cells or a battery, and it can be used in many studies as suggested below.

Experiments

- The ESP researcher can use this circuit to detect special abilities in a subject by variations in the amount of light falling onto a sensor. Filters in front of the sensor can be used to detect changes that occur only in specific light wavelengths (colors). The basic experiment is to monitor if, during ESP experiments, the amount of ambient light is affected or if the light reflected by the object on which the subject concentrates his mind changes (Fig. 162).
- Some experiments in psychokinesis are suggested by Fig. 163. First, the person can try to induce variations in light by directly affecting a source such as a common incandescent lamp. Or the person can induce variations in the amount of light passing through a medium such as a glass filled with a transparent or translucent solution, or even a solid piece of glass. Or the person can attempt to induce variations in the light reflected by a mirror.
- Any changes in ambient illumination can be detected if they occur in conjunction with a subject achieving the fourth state of consciousness or trance. The subject can be illuminated by a light source, with the device used to detect any change in the amount of light reflected by his body while in the trance or fourth state of consciousness. This is an important experiment to perform.

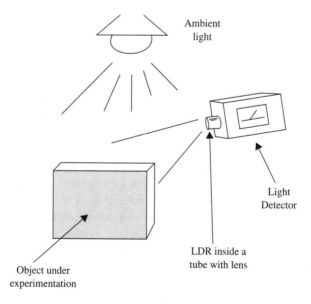

Figure 162 A paranormal experiment using the device.

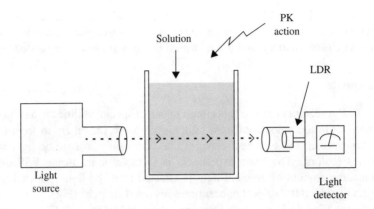

Figure 163 An experiment involving changes in the transparency of a solution.

- The manifestation of many paranormal phenomena are followed by changes in physical parameters in the ambient. For example, variations of the ac power line voltage can cause "flickering" of lamps, magnetic fields, and electric charges. To monitor these phenomena, an electroscope (Project 34) can be useful, along with a device that can detect small changes in the ambient light.
- The changes in the transparency of a solution with bacteria can be detected in a PK experiment or others.

How It Works

For the sensor, our circuit uses an LDR or photoresistor (CdS cell). This device has a resistance that changes with the amount of light falling onto the sensitive surface.

An LDR is more sensitive than a human eye, as it can detect amounts of light that are not detectable to us, and it also can detect light in part of the ultraviolet and the infrared ranges. The use of optical filters in front of the sensor (LDR), by making the circuit frequency selective, can help when working with light of one wavelength or frequency. It is also possible to add a polarization filter in front of the device.

The LDR is plugged to a Wheatstone bridge, where the null detector is a current meter. At a predetermined light level, the circuit is adjusted to the null point (zero current on the meter). Any change in the amount of light falling on the sensor changes the bridge's equilibrium, and a current flows through the meter. When the light level increases or decreases, this is represented as movement of the meter's needle in the corresponding direction. Using a meter with zero in the center of the scale, it is possible to detect very small changes of light intensity in either direction.

The circuit is powered from two AA cells, and the current drain is very small. The cells can last for many weeks or even months.

Assembly

Figure 164 shows the complete diagram of the light detector. As the circuit uses very few components, it can be mounted on a terminal strip.

The meter is a 50-0-50 μA type, but common types with scales ranging from 50 to 200 μA without zero in the center can be used. You only have to adjust P1 to a middle scale indication when using it.

The LDR or CdS cell can be of any type. It can be placed inside an opaque cardboard tube to increase sensitivity and directivity. A convergent lens in front of the device is another aid for increasing directivity and sensitivity. The LDR must be placed at the focal point of the lens.

All of the components can be installed in a small plastic or wooden box.

Testing and Using the Circuit

Place the cells in the cell holder. The on/off switch is not necessary.

Adjust P1 until you obtain a mid-scale indication on the meter. The LDR must be pointed to an illuminated spot. When you pass your hand in front of the sensor, the change in the light falling on it will cause the meter's needle to move.

When using the device, colored filters or polaroid filters can be placed in front of the sensor. Pieces of cellophane are good color filters that can be used in many experiments. Changes in the amount of light in an ambient can be detected by the movement of the meter needle.

Suggestions

- Two LDRs can be wired to the circuit to make a differential detector system. The circuit shown in Fig. 165 detects when the light levels in two different places change by different amounts.

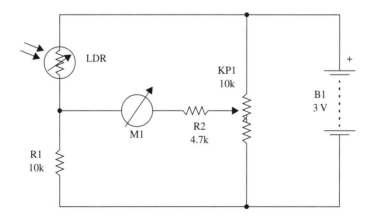

Figure 164 Light detector.

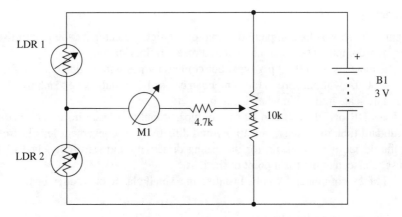

Figure 165 Differential version of the light detector.

- Add scale indications to the meter (+) and (–) to indicate when the amount of light increases or decreases.
- Change R1 according the amount of light used in the normal experiments. When working in bright ambients, use a 10 kΩ resistor. In dark ambients, increase this resistor to 100 kΩ.
- Use a resistance sensor to equilibrate the bridge from your skin resistance as shown in Fig. 166. This configuration can be used in experiments involving biofeedback, PK, and transcendental meditation.

Parts List: Project 35

Resistors

R1   10 kΩ, 1/8 W, 5%—brown, black, orange

R2   4.7 kΩ, 1/8 W, 5%—yellow, violet, red

Miscellaneous

P1   10 kΩ potentiometer

LDR Common light-dependent resistor or CdS cell (see text)

M1   50-0-50 μA meter (see text)

B1   3 V, two AA cells

Terminal strip, knob for P1, cell holder, plastic or wooden box, optical filters, wires, solder, etc.

Experimenting with Paranormal Skills

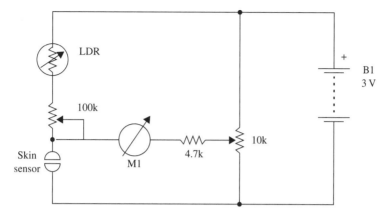

Figure 166 Light and skin resistance detector.

Project 36: Psycholamps

The effect of flickering lamps modulated by sounds sources can be used in many paranormal experiments. The circuit described here is a *strobe-rhythmic* lamp, i.e., a device that makes a common incandescent lamp flash and flicker in a frequency that changes according to the low-frequency signal applied to its input. White or colored lamps with power up to 100 W can be controlled by this circuit.

The sound can be produced by an amplifier that is fed from a variety of sources including microphones, to pick up the ambient sounds in biofeedback experiments, or a recording (CD or tape) with some relaxation sounds or special music. Some experiments for this circuit are suggested below.

Experiments

- The use of a stroboscopic light modulated by an audio source can help the subject in ESP experiments to be induced into trance state or to another appropriate mental state. The researcher can also verify how lamp flashes can interfere with the subject's ability to apply paranormal skills.
- In PK experiments, the experimenter can plug a microphone or other transducer into the input of the circuit and call on the subject to induce changes in the lamp's flash rate using his mind. The circuit also can be used to produce special mental states during the experiments.
- Some biofeedback and transcendental meditation experiments are also suggested. For instance, if we plug an oscillator that is controlled by skin resistance (as described earlier in this book) into the input of the circuit, the flash rate can be controlled by physiological changes in the subject's body. Using a special audio source such as percussion sounds records or white noise, the

flashes can be controlled to produce special states of mind, e.g., trances or the fourth state of consciousness. The frequency can be adjusted to closely match alpha rhythms in many experiments.
- For radiesthesia experiments, the researcher can compare the effect of a pendulum on a subject who is alternately exposed to the light of the stroboscopic source and not exposed to it. The source can help the subject achieve the trance state in experiments such as described before.
- Experiments involving the EIP and EVP can be performed using this light source. The circuit also can be used to induce a trance or other state of mind in sensitive subjects.

How It Works

The neon lamp, R1, P2, and C1 form a relaxation oscillator, driving the SCR. The frequency can be adjusted within a wide range of values by P2. Depending on the experiment, capacitor C1 can assume values in the range of 0.1 to 1 µF. Large values will lower the frequency.

The pulsed dc power necessary to operate this stage comes from the ac power line passing through D1.

The modulation stage begins with a transformer whose function is to pick up the audio signals from any external audio source, and also isolate it from the circuit. As common sources such as amplifiers have low-impedance, high-power outputs, the transformer must have a low-impedance winding as the input.

Resistor Rx is necessary to limit the power applied to the circuit, and its value is chosen according the amplifier or audio source power. The following table gives recommended values for this resistor.

Source Output Power (W)	Rx Value
0 to 1	—
1 to 5	10 Ω × 1 W
5 to 25	47 Ω × 1 W
25 to 50	100 Ω × 1 W
50 to 200	220 Ω × 2 W

According the output volume, P1 is adjusted to achieve the correct level of modulation. The signal is then applied to D2 and added to the pulses produced by the relaxation oscillator.

The SCR can control loads up to 3 A, but in this application we limited the loads to 100 W. See that the SCR is a half-wave control, and the lamp will flash with half of the total power. We can compensate for this by increasing the lamp power.

Assembly

Figure 167 is a schematic diagram of the psycholamp. The circuit is mounted on a printed circuit board as shown by Fig. 168.

The transformer is the noncritical component of this construction. Any transformer with a low-impedance or low-voltage winding and a high-voltage or high-impedance winding can be used experimentally. In particular, power supply transformers with primary windings rated to 117 Vac or even 220/240 Vac, secondaries rated at 5 to 12 V, and current ratings in the range between 50 and 500 mA can be used. The winding plugged into the input of the circuit (audio source) is the low-voltage one. The high-voltage winding is wired to P1.

The SCR is a TIC106, MCR106, or other equivalent in the "106" series. This component must have the suffix B or higher if the circuit is powered from the 117 Vac power line. If powered from the 220/240 Vac power line, the SCR must be rated for 40 V (Suffix D or higher).

The SCR must be mounted on a heatsink suited to the power of the controlled lamp. A piece of metal measuring 5 × 10 cm and bent to form a "U" is a suitable heatsink when working with loads up to 100 W.

The diodes also depend on the ac power line voltage. Use the 1N4004 for the 117 Vac power line or the 1N4007 for the 220/240 Vac power line.

In mounting the circuit, you must take care when installing it in a plastic or wooden box. The circuit is connected to the ac power line and can cause severe shocks if any live part is touched.

Although the circuit is powered from the ac power line, the audio source is isolated by the transformer and is no danger as long as its integrity is not compromised. **But the reader must be sure that the transformer used in this project is in perfect condition, with a high isolation resistance between windings.**

The external lamp can be plugged into the circuit using a common plug and a piece of wire, according to the location in which the reader wants to install it. Any common incandescent lamp with wattage ranging between 5 and 100 W can be used. The color is chosen according to the experiment.

The fuse is chosen according to the power of the controlled lamps. In an experiment, the reader can wire several lamps in parallel, since the total power does not exceed the limit of 3 A.

Testing and Using the Circuit

Connect the input of the device to the output of an audio amplifier shown in Fig. 169. Plug the power cord to the ac power line and turn on the amplifier, with any program source (a tape recording, a microphone, etc.) as input. Adjust the output volume to a comfortable level. It is not necessary to disconnect the loudspeaker from the output of the amplifier.

Set P1 to the minimum, and adjust P2 to obtain the desired lamp flash rate. Then, go to P1 and adjust this control until you can see changes in the flashes that

226 Part 3

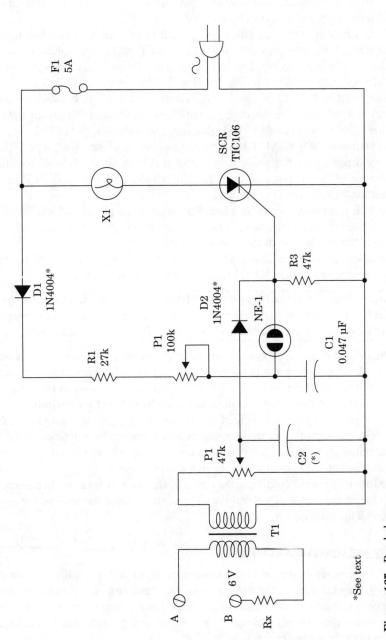

Figure 167 Psycholamp.

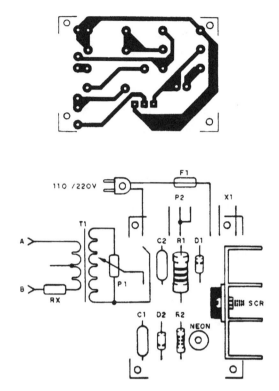

Figure 168 Printed circuit board for Project 36.

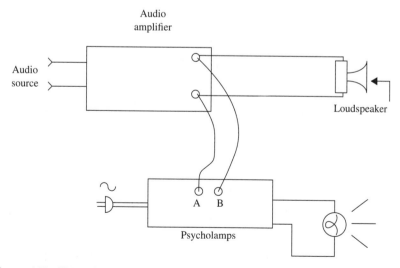

Figure 169 Using the psycholamp.

correspond to the variations in the sound level. At this point, the device is ready to be used. If a tape recorder is used as a sound source, you can plug it in from the monitor or phone output. In this case, resistor Rx is eliminated.

Suggestions

- Add a capacitor (C2) to change the frequency response of the device. If you include a high-value capacitor, the lamp will change in frequency only with the low-frequency sounds (bass).
- Add a switch in series with D1. Opening this switch, the circuit is converted into a rhythmic lamp. The lamp will flash only with changes in the sound level applied to the input.
- You can use a black light (ultraviolet) such as found in night clubs (incandescent type) to perform experiments with invisible light.

Parts List: Project 36

Semiconductors

SCR	TIC106B or TIC106D silicon-controlled rectifier
D1, D2	1N4004 or 1N4007 silicon rectifier diode

Resistors

R1	27 kΩ, 1/8 W, 5%—red, violet, orange
R2	22 kΩ, 1/8 W, 5%—red, red, orange
R3	47 kΩ, 1/8 W, 5%—yellow, violet, orange
Rx	According the output power of the audio amplifier (see text)

Capacitors

C1	0.47 µF/100 V, ceramic or metal film
C2	0.01 to 0.47 µF, ceramic or metal film (see text)

Miscellaneous

T1	Any transformer with a high-voltage or high-impedance winding and a low-impedance winding (see text)
P1	47 kΩ potentiometer
P2	100 kΩ potentiometer
NE-1	Any neon lamp (NE-2H or equivalent)
X1	Outlet to the lamp
F1	5 A fuse

Printed circuit board, heatsink for the SCR, knobs for the potentiometers, plastic or wooden box, power cord, fuse holder, wires, solder, etc.

Project 37: Electronic Candle

The mystical aspect of a flickering candle flame is an important element in many paranormal experiments or even rituals. When working with electronics in paranormal experiments, the aid of a circuit that can imitate the flickering effect of a candle, using a common incandescent lamp, can add some advantages in terms of final effects. One advantage is the absence of smoke and smell. The other is the danger of fire, which is avoided with the use of electronic resources. It is also worth noting that the researcher can use lamps of different color, and even IR or UV lamps.

The circuit described here imitates the flickering of a flame by controlling the brightness of a common incandescent lamp. Incandescent lamps with power in the range between 5 and 100 W or higher can be used in experiments, which is equivalent to the light produced by many candles. Many experiments can be suggested for using this circuit, as described below.

Experiments

- The flickering flame effect reproduced by a lamp can be useful to help a subject enter the trance state in an ESP experiment. It also can be used as a hypnotic aid in experiments involving transcendental meditation and biofeedback.
- The subject can try to change the light effect produced by this device using his mental powers. The effect can also be used to help the subject concentrate his mind on an object to be moved.
- Filling an ambient with the light produced by this device, the subject can achieve the fourth consciousness state or trance state more easily.
- In radiesthesia, the effects of an illumination that imitates a candle can be used in experiments involving the action of a pendulum and its effect on subjects.
- The mystic aspect of a candle illumination can be useful in experiments involving ghosts, spirits, and other paranormal manifestation. The location where the phenomenon will occur can be illuminated by this device.

How it Works

Two astable multivibrators using common transistors operate as modulators running in different frequencies. The first multivibrator has its frequency determined by C2 and C3 and the second by C4 and C5. The multivibrators use different values for these capacitors, as they have to produce square signals with duty cycles other than 50%. This means that short pulses applied to the lamp result in the flickering effect. The use of two oscillators also makes the lamp flicker at a random rate, as does a real candle.

The signals produced by the two multivibrators are combined by diodes D2 and D3 and applied to the gate of an SCR. This means that the SCR receives a random train of pulses, as shown in Fig. 170, producing the flickering effect.

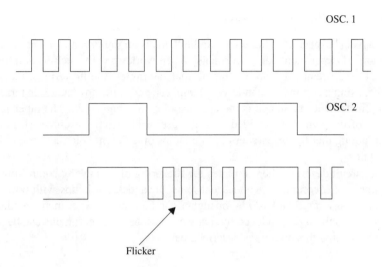

Figure 170 Signal in the SCR's gate.

The low voltage necessary to power the multivibrators comes from D1 and is filtered by C2. R1 has a value that depends on the ac power line. For a 117 Vac power line, a 10 kΩ resistor is indicated. A 22 kΩ resistor is recommended for a 220/240 Vac power line.

Note that all parts of the circuit, including the low-voltage stages, are connected to the ac power line. This means that the circuit must be assembled to avoid the possibility of anyone touching active parts to avoid shock.

Assembly

Figure 171 shows the complete schematic diagram of the electronic candle. The components are placed on a printed circuit board as shown in Fig. 172.

The SCR must be mounted on a small heatsink. The suffix depends on the ac power line voltage. If the circuit will be powered from a 117 Vac power line, types with B or D suffixes are suitable. For a 220/240 Vac power line, use the types with suffix D.

Any incandescent lamp rated in the range between 5 and 100 W can be used. Types that look like a candle are recommended if the reader can find them.

The transistors are not critical, and the electrolytic capacitors should have voltage ratings to 16 V or more.

As the circuit is powered from the ac power line and no isolation transformer is used, special care must be taken to avoid shock. Isolate the circuit inside a plastic or wooden box.

The lamp can be mounted on the upper side of the box or, if you prefer, placed at a distance from the device and connected by a long wire.

Experimenting with Paranormal Skills 231

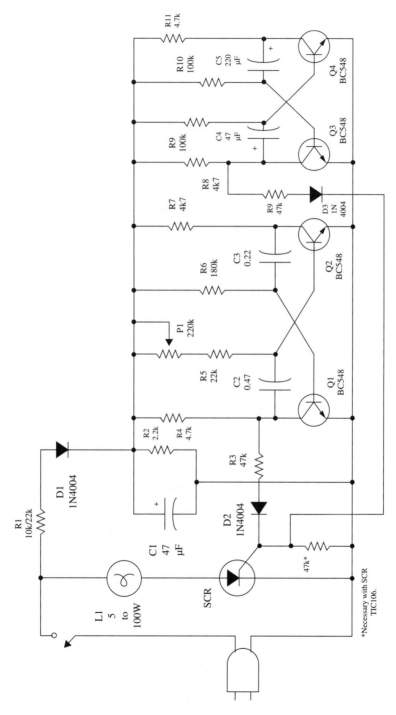

Figure 171 Electronic candle.

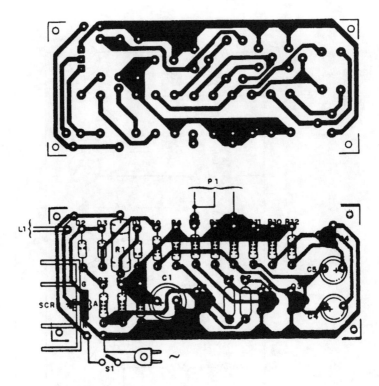

Figure 172 Printed circuit board for Project 37.

Testing and Using the Circuit

Plug the device into any ac power line and turn on S1. Adjust P1 to have a flickering effect in the lamp to suit your expectations. If you don't like the effects (e.g., if, due to component tolerances, they do not suitably imitate a candle), you may want to make some alterations in the circuit. See in the item in the "Suggestions" section for instructions on how to do that.

If the circuit doesn't function, check the voltage at C1. The voltage must be 8 V. If necessary, alter the value of R2 to achieve this voltage.

Suggestions

- To change the flickering effect, you can make the following component changes:
    C2 in the range between 0.22 and 1 µF
    C3 in the range between 0.047 and 0.47 µF
    C4 in the range between 10 and 220 µF
    C5 in the range between 47 and 470 µF

These alterations can compensate for tolerance problems in the capacitors.

- Use an IR or UV incandescent lamp for the experiments.
- Replace R11 with an LDR and control the effect by the ambient light. The LDR must not be illuminated by the device's circuit lamp, or you will get feedback.
- You can also replace R6 with an LDR to control the effect in another point of the circuit.

Parts List: Project 37

Semiconductors

SCR	MCR106 (4 or 6) or TIC106-B or D (according the AC power line) silicon-controlled rectifier (see text)
Q1, Q2, Q3, Q4	BC548 or equivalent - general purpose NPN silicon transistors
D1, D2, D3	1N4004 or 1N4007 - silicon rectifier diodes

Resistors

R1	10 kΩ (117 Vac) or 22 kΩ (220/240 Vac) × 10 W, wire-wound (see text)
R2	2.2 kΩ, 1/8 W, 5%—red, red, red
R3, R9	47 kΩ, 1/8W, 5%—yellow, violet, orange
R4, R7, R8, R12	4.7 kΩ, 1/8W, 5%—yellow, violet, red
R5	22 kΩ, 1/8W, 5%—red, red, orange
R6	180 kΩ, 1/8 W, 5%—brown, gray, yellow
R10, R11	100 kΩ, 1/8 W, 5%—brown, black, yellow

Capacitors

C1, C4	47 µF/16 WVDC, electrolytic
C2	0.47 µF, ceramic or metal film
C3	0.22 µF, ceramic or metal film
C5	220 µF/16 WVDC, electrolytic

Miscellaneous

P1	47,000 to 220 kΩ potentiometer
S1	SPST, toggle or slide switch
X1	Incandescent lamp, 5 to 100 W (see text)

Printed circuit board, plastic or wooden box, power cord, heatsink for the SCR, knob for the potentiometer, wires, solder, etc.

3.7 ESP and PK Experiments

> *Nihil est in intellectu quod non prius fuerit in sensu.*
> *(Nothing is in the intellect that was not passed before by the senses.)*
>
> Unknown Latin author

Extrasensory perception and psychokinesis or telekinesis form one of the most important fields in paranormal research. The aim of these two phenomena lies in the powers of the mind. Can a person read others' minds? Can a person know what is happening in other places? Can a person move an object on a table using his mind powers?

To conduct research with these paranormal phenomena, the reader can be assisted by several electronic devices. The space we have in this book is insufficient to describe all the devices that are suitable for experiments in this field. Therefore, we have selected some that are simple to build and easy to use in practical experiments.

The circuits intended for ESP are random number generators. In experiments in this area, as described before, researchers may use the electronic devices for various purposes.

Project 38: Random Number Generator

A random number generator is an important device for use in experiments involving ESP. Zener cards can be affected by either the subject or the researcher when they are handled. Dies and coins can also be influenced by the researcher or the subject in an experiment. But we can consider an electronic device that chooses random numbers to be a subject-validation device, as it can be influenced by the subject when it displays a number or card but not by the researcher.

The circuit described here is ideal to perform both ESP experiments and PK as, in the first case, the subject must use his mind to guess what number has been chosen by the circuit and, in the second case, he must try to predict what number is about to be chosen by the circuit.

A principal advantage of this circuit is that it is powered from a 9 V battery and can be used in any convenient location. The circuit can be used in several experiments as suggested below.

Experiments

- The researcher can use the device to choose a number that he does not yet know and make tests with subjects who try to guess the number using ESP.
- The subject can try to control what number is chosen by the circuit in psychokinesis experiments.
- In radiesthesia experiments, the device can be used to determine if the pendulum activity can alter the numbers chosen by the circuit.

How It Works

Q1, Q2, and Q3 form an oscillator that is programmed to produce a random number of pulses when S is pressed and released. The secret of this circuit is that it does not stop the circuit when S is released, avoiding the influence of the subject in the selected number of pulses. When S is released, the oscillator's frequency slows down until it stops some seconds later. The time span between the instant in which S is released and the circuit stops presenting random numbers is determined by C1 and R1. These components can be changed to suit the experiments.

Now, let us see in detail how this stage operates. When S1 is pressed, C1 charges with the power supply voltage biasing Q1. Q1 controls the relaxation oscillator formed by a unijunction transistor (UJT). When biased at the moment in which S is pressed, Q1 starts with the UJT oscillation at a frequency adjusted by R1 and basically determined by C2. These components determine how fast the numbers will run in the sorting process.

The pulses produced by the UJT are amplified by Q3, driving a decade counter formed by the CMOS IC 4017. In this application, the counter is programmed to count up to six, meaning that the device can generate one of six numbers. If necessary, the reader can alter the circuit to count up to other numbers (between 2 and 10).

When receiving the pulses, the outputs of the 4017 go to the high logic level in sequence, driving transistors Q4 to Q9. As a load, these transistor use LEDs, but the circuit can be altered to drive small incandescent lamps (up to 50 mA).

When the last output goes to the high logic level, the circuit is reset and, in the next pulse, the first output again comes to the high logic level, driving the corresponding LED. This means that, as high as the number of pulses can be, there is always one LED remaining high after the cycle.

To avoid stopping the circuit when S is released, we have the capacitor C1 in the circuit. When S is released, C2 needs several seconds to discharge through the transistor, maintaining it in conduction, and so the oscillator remains on. As long as the capacitor discharges, the resistance between the emitter and collector of Q1 increases, and then the frequency of the oscillator falls. The final effect is the gradual reduction of the running speed of the LEDs until the moment in which only one of them remains on. This corresponds to the generated number. To start the process again, the reader has only to press and release S again.

Assembly

Figure 173 shows the random number generator's diagram. The components are placed on a small printed circuit board as shown in Fig. 174.

Common red LEDs are used. Keep the terminals long, as shown in Fig. 175. This way you can fix the printed circuit board in the box and the LEDs will appear in the holes in the panel.

It is important to observe the position of polarized components such as the LEDs, electrolytic capacitors, bipolar and UJT transistors, and battery. The other

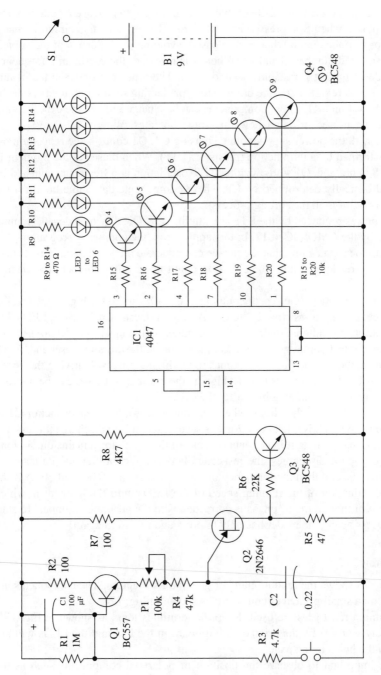

Figure 173 Random number generator.

Experimenting with Paranormal Skills   237

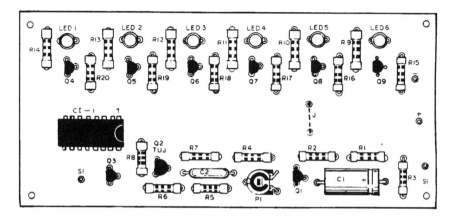

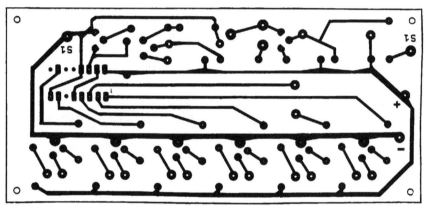

Figure 174   Printed circuit board for Project 38.

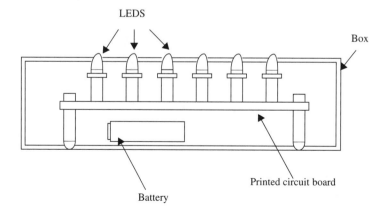

Figure 175   Installing the printed circuit board in a box.

components are not critical, and small changes in the original values will not affect the performance of the circuit. All of the components are placed inside a plastic or wooden box.

You can devise cardboard removable panels to be used in conjunction with the LEDs for various experiments. For example, zener cards, playing card symbols, numbers, letters, or other symbols can be used as shown in Fig. 176.

Testing and Using the Circuit

Connect a battery to the battery clip and press S. The LEDs will run at a speed that can be adjusted by P1. Make sure all LEDs glow in the process. Release S and wait a few seconds until one LED remains on.

The circuit is now ready to be used. You can conduct ESP experiments in many forms.

1. Generate a number without the subject knowing it, then ask him to write the number on a sheet of paper or guess it aloud.
2. Let the subject press S to generate the number, but cover the LEDs so they cannot be seen. After the subject tries to guess the number verbally, uncover the LEDs. Note the number of correct answers. S can be placed far from the subject to make sure that the number is hidden.
3. In a PK experiment, the subject presses S and releases it. He then tries to use his mental powers to dictate the generated number.

Suggestions

- Use IR LEDs in ESP experiments. An IR detector can be used to see which LED is on during the experiment.

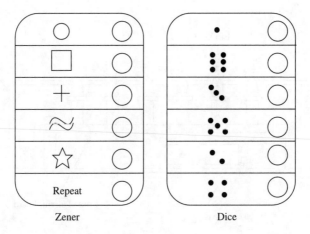

Figure 176 Cards to be placed in the box.

- Alter C1 to change the length of the number generation process. Greater values increase the time. Values between 22 µF and 1,000 µF can be used experimentally.
- Alter C2 to change the speed of the LEDs. Smaller capacitors will increase the speed. Values between 0.047 and 1 µF can be tested.
- In PK or ESP experiments, resistor R1 can be increased to 22 kΩ, C1 can be reduced to 0.1 µF, and you can replace S with a touch sensor. The circuit then can be controlled by skin resistance or by a plant attached to the sensor.

Parts List: Project 38

Semiconductors

IC1	4017 CMOS integrated circuit
Q1	BC557 or equivalent general-purpose PNP silicon transistor
Q2	2N2646 unijunction transistor
Q3 to Q9	BC548 or equivalent general-purpose NPN silicon transistors
LED1–LED6	Red common LEDs (see text)

Resistors

R1	1 kΩ, 1/8 W, 5%—brown, black, green
R2, R7	100 Ω, 1/8 W, 5%—brown, black, green
R3, R8	4.7 kΩ, 1/8 W, 5%—yellow, violet, red
R4	47 kΩ, 1/8 W, 5%—yellow, violet, orange
R5	47 Ω, 1/8 W, 5%—yellow, violet, black
R6	22 kΩ, 1/8 W, 5%—red, red, orange
R9–R14	470 Ω, 1/8 W, 5%—yellow, violet, brown
R15–R20	10 kΩ, 1/8 W, 5%—brown, black, orange

Capacitors

C1	100 µF/12 WVDC, electrolytic
C2	0.22 µF, ceramic or metal film

Miscellaneous

P1	100 kΩ, trimmer potentiometer
S	Momentary contact, normally open switch
S1	SPST, toggle or slide switch
B1	9V battery

Printed circuit board, battery clip, plastic or wooden box, wires, solder, etc.

Project 39: Binary Random Number Generator

This circuit generates a number between 0 and 15 in a binary form. The numbers will appear with values between 0000 and 1111, given by the combination of four LEDs according to the table below.

Number	LED1	LED2	LED3	LED4
0	0	0	0	0
1	0	0	0	1
2	0	0	1	0
3	0	0	1	1
4	0	1	0	0
5	0	1	0	1
6	0	1	1	0
7	0	1	1	1
8	1	0	0	0
9	1	0	0	1
10	1	0	1	0
11	1	0	1	1
12	1	1	0	0
13	1	1	0	1
14	1	1	1	0
15	1	1	1	1

The numbers are given by four LEDs, each of which can represent 0 (off) and 1 (on). They can be on or off in any combination as shown in the table.

When you press S, the LEDs will flash to produce random combinations. When S is released, the LEDs will remain in one of the combinations correspondent to the generated number. Unlike the previous project, this circuit does not have an "inertia" circuit that causes a delay in arriving at the generated number.

The circuit is powered from AA cells and can be used in basically the same experiments suggested for Project 38.

Experiments

- The subject must guess what number will be generated. As suggested in the previous project, there are several ways to improve the tests. The LEDs can be

associated with combined elements such as number and color, or a table with 16 different symbols can be created.
- As in the previous project, the subject can perform PK experiments by trying to set the generated number using his mental powers. See the next project for more suggestions about experiments of this type.
- In transcendental meditation and biofeedback, the LEDs running in many combinations can help the subject to reach the necessary mental states to make experiments in this area. The speed of the LEDs can be reduced, as indicated in the "Suggestions" section, to help the experimenter if appropriate.
- For radiesthesia experiments, the influence of a pendulum on the results can be tested.
- *UFOs and ghosts:* How can ghosts be detected by a generated number? Can the presence of paranormal phenomena in a particular location influence the numbers generated by this circuit? You can experiment to find out.

How It Works

Two of the four NAND gates of a TTL IC 7400 are used as an oscillator whose frequency is determined by R1 and C1. This circuit applies its signal to a third NAND gate, controlled by S1 in a manner such that pulses pass through it only if S1 is pressed. This means that, when pressing S1, the pulses generated by the oscillator are applied to the second block of the circuit, formed by IC2.

IC2 is a divider by 2 and divider by 8 in a single chip. This means that it is a binary divider that can count up to 16 and give the results in four outputs, Q1, Q2, Q3, and Q4.

To each output is connected an LED that indicates its state. The LED will glow when the output is high, representing a 1, and will remain off if the output logic level is a 0.

As with all TTL circuits, the power supply voltage must be 5.0 V. In practice, the circuit will function with any voltage between 4.5 and 5.5 V. This means that four AA cells are *not* a suitable power supply for this circuit. To power the circuit from AA cells, we have to add two common diodes in series with them as shown in Fig. 177.

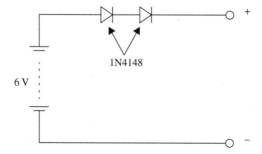

Figure 177 Powering the circuit from AA cells.

242   Part 3

Each diode will produce a voltage drop of about 0.6 V in the power supply voltage, reducing the total voltage applied to the circuit to 4.8 V. This is a suitable value for its operation.

You can also power the circuit using the power supply shown in Fig. 178. The input transformer has a primary winding rated for the ac power line (117 Vac), a secondary winding of 7.5 to 9.0 V, and currents in the range between 50 and 300 mA.

The diodes have a voltage rating of 12 V or more. It is not necessary to place the IC 7805 on a heatsink, as the circuit drains a very low current.

Assembly

Figure 179 shows the complete diagram of the binary random number generator. The components can be placed on a small printed circuit board as shown in Fig. 180.

Observe the position of the polarized components, such as the LEDs. If they are reversed, the circuit will not function properly. The LEDs can be mounted with long terminals to allow them to pass through holes in the panel as described in the previous project.

The circuit can be installed in a plastic or wooden box. The size will depend on the battery holder format and size or if the assembler intends to use a power supply with a transformer.

The panel can be designed to use cardboard panels with symbols as described in the previous project.

Testing and Using the Circuit

Place the cells in the cell holder and press S1. The LEDs will flash at a fast rate. In some cases, you will have the sensation that the LED connected to Q1 and Q2 glows with low brightness because of its flash speed. When you release S1, the LEDs will stop in one of the combinations shown in the table. Try it a few more

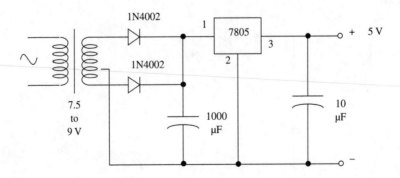

Figure 178   Power supply for Project 39.

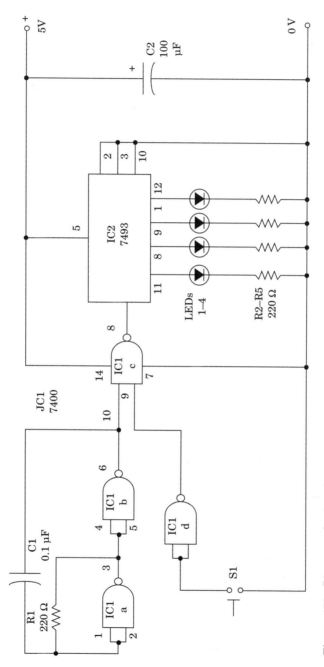

Figure 179 Binary random number generator.

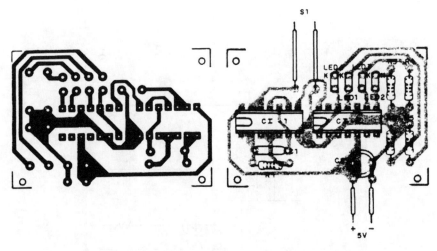

Figure 180 Printed circuit board for Project 39.

times to make sure the next combination isn't the same. If the circuit doesn't run, try using one diode instead of two in the power supply (if powered by cells).

Using the Circuit

The experiments for this device are the same as described for the previous project.

Suggestions

- You can make the LEDs run faster by reducing the value of C1. Values between 0.01 and 0.044 µF can be used.
- You can add inertia to the circuit to make the LEDs run for a few seconds after S1 is released. This is accomplished by installing a 100 to 470 µF capacitor in parallel with S1.

Parts List: Project 39	
Semiconductors	
IC1	7400 TTL IC, four two-input NAND gates
IC2	7493 TTL IC, dual flip-flop
LEDs 1–4	Common red LEDs
Resistors	
R1	220 Ω, 1/8 W, 5%—red, red, brown
R2–R5	330 Ω, 1/8 W, 5%—orange, orange, brown

Capacitors

C1   0.1 µF, ceramic or metal film

C2   100 µF/6 WVDC, electrolytic

Miscellaneous

S1   Pushbutton, normally open

B1   Four AA cells + two diodes (1N4148) or a 5 V power supply

Printed circuit board, plastic or wooden box, wires, battery holder, solder, etc.

3.8 UFOs and Ghosts

> *Most men become convinced that they are right only after other people agree with them. But some of us find nothing more unsettling than our own words on the lips of others.*
>
> Walter Kaufmann

UFOs and ghosts are important paranormal phenomena to be studied with electronic equipment. The presence of these subjects in any location causes some physical state alterations that can be detected with electronic devices.

Aside from alterations in electric fields, the most common variations related to phenomena involving ghosts and UFOs are changes in ambient magnetic fields. Such changes commonly cause performance deviations in devices such as watches, TV sets, electrical circuits (e.g., auto ignitions), and many others.

However, in many cases, the variations are very weak, their effects do not appear in common appliances, and no other visible modification in the ambient can help the researcher to detect the phenomena using only his senses. In such cases, electronics devices are important. Accordingly, the next projects deal with the detection of magnetic fields.

Project 40: UFO Detector

Changes in magnetic fields are among the physical manifestations associated with the presence of a UFO in a nearby location. TV sets and other electronic and electrical circuits can suffer interference to a degree that depends largely on how far the UFO is from them. If the UFO is not very close, the magnetic field changes will not be strong enough to cause any visible change in these devices.

One way to detect a UFO, even at a considerable distance, is through the use of a magnetic field detector or UFO detector such as the one described here. This circuit detects small changes in the ambient magnetic field, which trigger an alarm tone that lasts for a few seconds.

The circuit is sensitive enough to be triggered by some natural magnetic field changes (e.g., lightning strikes during a storm or even when a nearby electrical device is turned on). However, when used in field studies, distant from any source of interference and on a clear day (or night), a triggering event can be a strong signal that a UFO is nearby.

Experiments

- This device is useful not only for UFO detection experiments. In ESP experiments, the circuit can detect whether any change in the ambient magnetic field is produced when a subject is in a trance state in experiencing ESP.
- A subject can try to control or alter the ambient magnetic field using mental powers in PK experiments.
- As in ESP experiments, the subject can try to control the ambient magnetic field in transcendental meditation experiments or for biofeedback.

How It Works

Any change in the ambient magnetic field produces a small voltage in the coil's terminals. This voltage is applied to the inputs of an operational amplifier (IC1). The voltage gain of this amplifier is determined by the feedback network formed by P1. The higher the resistance of this component, the higher the voltage gain of the circuit. The amplified voltage is then applied to the base of Q1, which forms a second amplification stage.

When this transistor is saturated with a signal coming from the previous stage, the trigger input of a 555 timer IC, wired as a monostable multivibrator, goes to the low logic level. At this moment, the monostable triggers, with its output going to the high logic level.

The output will remain in the high logic level even when the detected magnetic field changes disappear. The time span during which the output remains high depends on R5 and C2.

The next stage, triggered by the monostable multivibrator, is a two-transistor audio oscillator. This oscillator drives a loudspeaker, and its frequency is determined by R6 and C3.

Power comes from four AA cells (6 V), and current drain is very low, extending battery life.

Assembly

Figure 181 shows a diagram of the UFO detector. The components are placed on a small printed circuit board as shown in Fig. 182.

The coil is the critical component in this project. Sensitivity depends on the number of turns on the coil. The higher the number of turns, the more sensitive the device.

Experimenting with Paranormal Skills 247

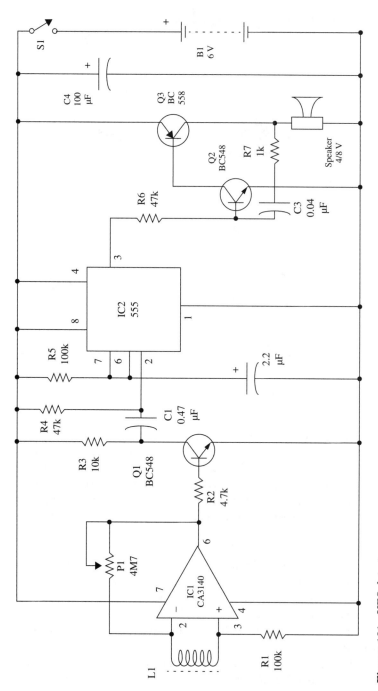

Figure 181 UFO detector.

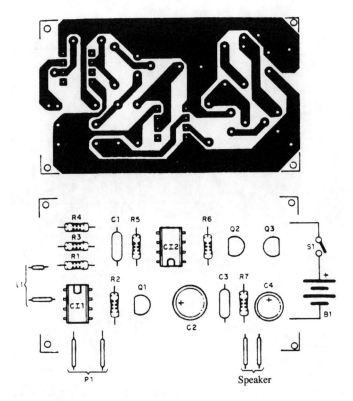

Figure 182 Printed circuit board for Project 40.

The reader has two possibilities to obtain this component. One is to wind 5,000 or more turns of any wire between 30 and 34 AWG on a ferrite rod (any size that can accept that many turns). The other is use the primary winding of any transformer with a high-voltage or high-impedance transformer. The primary winding of a 117 Vac × 6 V (50 to 250 mA) transformer can be used as the pickup coil. You only have to take out the core, leaving only the form with the wire as shown in Fig. 183.

A ferrite rod is then placed in the core (you can glue it or hold it in place using paper or a plastic sponge to fill the empty space). Ferrite rods with diameters between 0.8 and 1.2 cm, and lengths in the range of 10 to 20 cm, can be used.

The circuit can be installed into a small plastic or wooden box. The dimensions of the box basically depend on the size of the loudspeaker and the cell holder.

If the ferrite rod is too long to fit inside the box, make sure that both ends extend outside the box.

On the front panel we have two controls: the on/off switch and the sensitivity control. When mounting the circuit, observe the position of polarized components such as the transistors and the electrolytic capacitors.

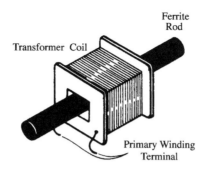

Figure 183 Detail of X1.

Testing and Using the Circuit

Place the cells in the cell holder and turn on the power supply (close S1). Open P1 to put the unit in the higher gain condition.

Quickly pass a small magnet near the ferrite core. The circuit will be triggered, producing a tone for a few seconds.

You can change the tone by adjusting C1 or R6. The length of time that the tone is on is determined by R5. Values between 22 kΩ (for short tones) and 470 kΩ can be tested.

In use, place the device in any suitable location (i.e., at a distance from ac power lines or devices that can trigger it, such as cars, telephone lines, etc.) and turn it on. The production of any tone indicates the presence of a magnetic field change in that location.

Note: the circuit tends to be triggered by atmospheric electrical discharges such as lightning from an approaching storm.

Suggestions

- You can increase the unit's sensitivity by adding a 10 MΩ resistor in series with P1.
- Fast changes in the magnetic field can become undetectable if a capacitor with values in the range between 1 and 10 μF is placed between the base of Q1 and the 0 V power line.
- The tone produced by the oscillator can be adjusted by replacing C3 with one of a different value, between 0.01 and 0.1 μF.
- A tone control for the signal can be added. Replace R6 with a 10 kΩ resistor and add a 100,000 Ω trimpot in series with this component.
- A photo diode can replace the coil. In this configuration, the circuit will detect small changes in light intensity or light flashes in an ambient.

Parts List: Project 40

Semiconductors

IC1	CA3140 operational amplifier, integrated circuit
IC2	555 timer, integrated circuit
Q1, Q2	BC548 or equivalent general-purpose NPN silicon transistor
Q3	BC558 or equivalent general-purpose PNP silicon transistor

Resistors

R1, R5	100 kΩ, 1/8 W, 5%—brown, black yellow
R2	4.7 kΩ, 1/8 W, 5%—yellow, violet, red
R3	10 kΩ, 1/8 W, 5%—brown, black, orange
R4, R6	47 kΩ, 1/8 W, 5%—yellow, violet, orange
R7	1 kΩ, 1/8 W, 5%—brown, black, red

Capacitors

C1	0.47 µF, ceramic or metal film
C2	2.2 µF/16 WVDC, electrolytic
C3	0.047 µF, ceramic or metal film
C4	100 µF/12 WVDC, electrolytic

Miscellaneous

L1	Pickup coil (see text)
P1	4,700 kΩ potentiometer
SPKR	5 cm x 4/8 Ω loudspeaker
S1	SPST, toggle or slide switch
B1	6 V, four AA cells

Printed circuit board, plastic or wooden box, cell holder, knob for P1, wires, solder, etc.

Project 41: Ghost Finder

Paranormal phenomena are not only detectable by our common five senses; they can be manifested in ways that our senses cannot pick up. They can produce variations in magnetic fields, electric fields, and the invisible light spectrum. The "ghost buster" or paranormal researcher must enlist the aid of equipment that detects changes in physical quantities that are beyond the reach of our senses.

The interesting device described here is a magnetic field converter that can distinguish changes in a magnetic field or pick up modulated magnetic fields and convert them into audible sounds. We might say that this is a device to "hear magnetic fields," and many experiments in paranormal sciences can be devised with its aid.

The circuit is formed by a pickup coil and amplifier stages that drive a small loudspeaker. It is very simple and can be powered from AA cells, making it portable.

Experiments

In ESP experiments, the device can be used to monitor any abnormal signals present near the subject. The magnetic changes in the ambient when the subject reaches the trance state or other mind states can be reproduced as audible sounds.

This type of bioprobe can be used to detect whether, in the location of the experiments, there exist magnetic fields or signals that can cause interference. This is useful in locations where transcendental meditation or biofeedback experiments are made, and in many other places. In radiesthesia experiments, we can monitor the influence of a pendulum on the ambient magnetic field during an experiment.

According to many studies, ghost and spirit manifestations often are associated with magnetic field disturbances. Such disturbances can be detected using this circuit. In EVP experiments, the circuit can also be used to pick up signals from a telephone handset via its magnetic field.

How It Works

The modulated magnetic field is picked up by the coil, producing a small signal at its output. This signal is applied to the inputs of a JFET operational amplifier wired in the differential mode. The voltage gain of this configuration is regulated by the feedback potentiometer, which can be adjusted from 1 to 47 MΩ.

In the next stage, we find an audio amplifier that can drive a small loudspeaker using the several hundred milliwatts of signal coming from the first stage. As the voltage gain is given by the feedback network in the operational amplifier, the volume control has been replaced with a 10 kΩ fixed resistor. The circuit is powered by four AA cells.

Assembly

Figure 184 shows a complete diagram of the device. The components are mounted on a small printed circuit board as shown in Fig. 185.

The pickup coil is the same as described in the previous project. You can also use the primary winding of a small high-voltage or high-impedance transformer without the metallic core.

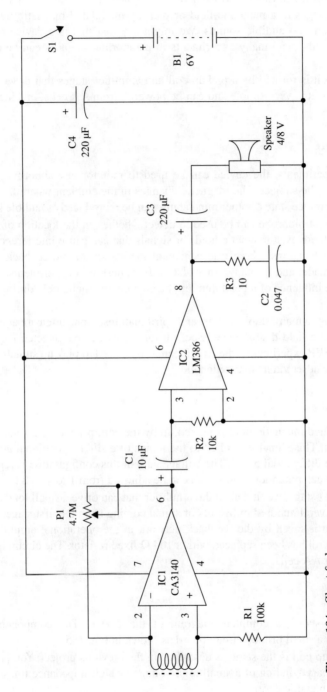

Figure 184 Ghost finder.

Experimenting with Paranormal Skills 253

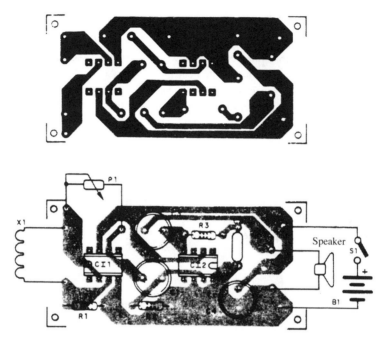

Figure 185 Printed circuit board for Project 41.

A small plastic or wooden box can be used to house all the components. Its size is basically determined by the dimensions of the loudspeaker and the battery holder.

The pickup coil can be installed at the end of a cable as shown in Fig. 186. The cable must be shielded and about 1 to 1.5 m in length. Other components used in the project are not critical, and many of them can be replaced with equivalents that have values near the recommended ones.

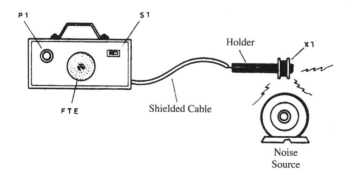

Figure 186 Using the ghost finder.

Testing and Using the Circuit

Place the cells in the cell holder and turn on S1. Placing the coil near magnetic field sources such as those plugged to ac power lines (motors, solenoids, etc.), the "hum" of the power supply line can be heard.

If the pickup coil is placed near a loudspeaker that is reproducing any sound, the signal will be picked up from its magnetic field (rather than from vibrations in the air, as in the case of a microphone) and reproduced in the loudspeaker of the unit.

Using this circuit is very simple. You only have to place the coil near the source of the magnetic disturbances you want to detect. The gain is adjusted by P1. Avoid locations where the hum of the ac power line can be picked up. Changing the position of the coil, you can find areas where the field lines cut the coil wire at a square angle. In this position, no voltage is induced, and the hum will not interfere with the experiments.

Suggestions

- Replace R2 with a 10 kΩ potentiometer to add a volume control to the circuit. One end is connected to pin 3, the other to capacitor C1, and the cursor to pin 3 of the IC.
- More gain can be added to the circuit if you replace P1 with a 22 MΩ resistor.
- The tone can be controlled with a capacitor placed between pin 3 and the 0 V line. Values for this capacitor can range from 0.01 to 0.22 µF. The higher the capacitor value, the more you can cut the high frequencies (treble).
- A photodiode or phototransistor can replace X1. The circuit is thereby transformed into a light-to-sound converter.
- Placing this circuit near the handset of a telephone, the sounds on the line can be picked up and reproduced by the loudspeaker. Experiments involving sounds from the telephone line can be performed.

Parts List: Project 41

Semiconductors

IC1   CA3140 JFET operational amplifier, integrated circuit

IC2   LM386 audio amplifier, integrated circuit

Resistors

R1   100 kΩ, 1/8 W, 5%—brown, black, yellow

R2   10 kΩ, 1/8 W, 5%—brown, black, orange

R3   10 Ω, 1/8 W, 5%—brown, black, black

Capacitors

C1	10 µF/12 WVDC, electrolytic
C2	0.047 µF, ceramic or metal film
C3	220 µF/12 WVDC, electrolytic
C4	100 µF/12 WVDC, electrolytic

Miscellaneous

P1	4,700 kΩ potentiometer
X1	Pickup coil (see text)
SPKR	4/8 Ω, 5 cm small loudspeaker
S1	SPST, toggle or slide switch
B1	6 V, four AA cells

Printed circuit board, cell holder, plastic or wooden box, knob for P1, wires, solder, etc.

3.9 Other Paranormal Experiments

In this book we have described only a small part of all the experiments the reader can design using simple electronic circuits. Many of the projects described herein are experimental and simple and can be replaced by extended versions, including some commercial units. For instance, the magnetic field detector described in Project 40 can be replaced by a commercial amplifier with very high gain or used with the low-impedance preamplifier also described in this book.

The use of our circuits in initial experiments is important to show how paranormal phenomena can be detected or produced. It is up to the reader to continue with more advanced equipment or even upgrade the circuits shown here. Combining some of the circuits described herein, the reader can devise many experiments in addition to the ones suggested here.

A list of new experiments that combine our projects is given below. The reader is free to alter them or add new elements to these experiments. We call the reader's attention again to the fact that imagination is fundamental to discovering new things in the paranormal sciences (or in any other field of research).

You can combine appropriate devices to perform the following experiments:

- Sound and light can be used in biofeedback experiments, adding light effects such as the electronic candle.
- Monitor effects in biofeedback and TM experiments by simultaneously using the electroscope, a magnetic field detector, and a light detector.
- Use light and sound effects to induce trances and other mental states in your experiments. Combine the devices in various ways.
- Use biofeedback devices in ESP experiments with random number generators.

- Combine thermal detectors and skin resistance change detectors in experiments with sensitive subjects.
- In EIP and EVP experiments, use such light effects as the candle and the strobo-rhythmic device to induce special conditions in an ambient.
- Combine light effects (e.g., the candle and strobo-rhythmic device) with sound effects (ultrasonic generators and others) in experiments involving ghost detection and radiesthesia.

3.10 The Computer

The computer (PC) is a powerful tool for use in any field of research, and this includes paranormal sciences. Although we have included suggestions for the use of a PC in only a few cases herein, the paranormal researcher can incorporate its processing power in nearly all of the experiments described in this book. The computer is particularly powerful for processing large amounts of data collected during ESP trials, where statistics form the basis of all of the experiments.

A computer can also be used to pick up information directly from circuits or transducers. This can be accomplished with the aid of an analog-to-digital (A/D) converter. An A/D converter is a circuit that converts analog signals (e.g., voltage or resistance between two electrodes) into digital data that can be processed by a computer. This is illustrated in Fig. 187. The data are applied as 8 or 16 bit information, which provides a resolution of 256 or 64,000 steps or values.

A reader who is experienced with data acquisition software can program the PC to pick up the data from the input at regular intervals or at a predetermined time and convert them into graphics, show them in tabular form, or use them for complex calculations. Many programs are suitable for use with the data supplied by A/D converters that are plugged into the PC's parallel port.

Among the most suitable for this task are Delphi and Visual Basic (V/B), but an experienced reader can use programs written in other languages, including Java, Q-BASIC, and even Debug.

Figure 188 shows a simple circuit to convert analog data into digital data to be plugged to the input of a PC. This circuit can be mounted on a printed circuit board as shown in Fig. 189.

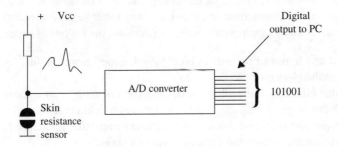

Figure 187 Using an A/D converter.

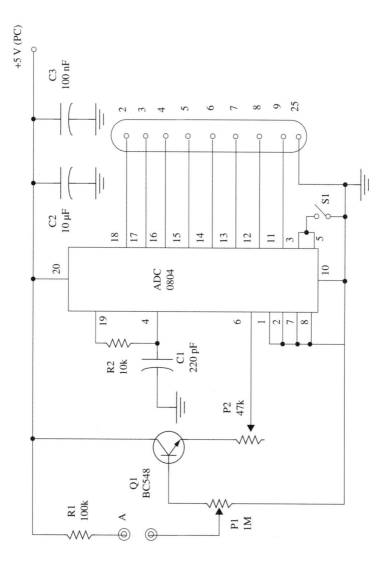

Figure 188 An A/D converter for resistive transducers.

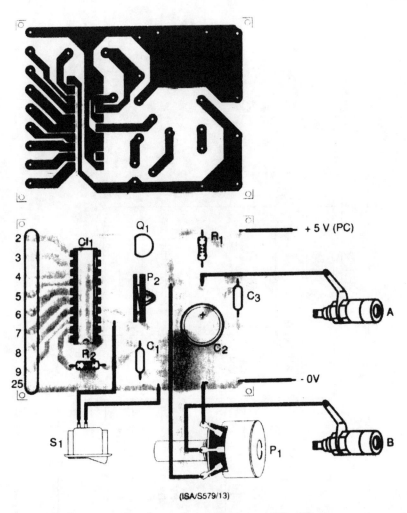

Figure 189 Printed circuit board for the circuit shown in Fig. 188.

The numbers in the output correspond to the pins of the DB25 connector used to transfer the signals to the parallel I/O port of the PC. Note that the circuit must be powered from a regulated 5 V power supply.

It is possible that, in a second volume in this series, we will concoct more projects that include advanced applications of A/D converters for paranormal experiments.

Bibliography

1. Rolf H. Krauss, *Beyond Light and Shadow.*
2. Cyril Permutt, *Beyond the Spectrum.*
3. Dr. Konstantin Raudive, *Breakthrough.*
4. Newton C. Braga, *CMOS Projects for the Experimenter,* Newnes, 1999.
5. Newton C. Braga, *Curso Prático de Eletrônica (in Portuguese),* Editora Saber, 1998.
6. C. Bertelsann, *Das UFO Phnomen Munique,* 1978.
7. *Eletrônica Total,* many editions from number 1 to 70, *Brazilian Electronics Magazine* (in Portuguese).
8. Kendrick Frazier, *Encounters with the Paranormal.*
9. Joseph Banks Rhine, *Extra-Sensory Perception,* Boston, Bruce Humphries, 1934.
10. Newton C. Braga, *Fun Projects for the Experimenter,* Prompt Publications, 1998.
11. Robert Laffont, *Le Livre Noir des Soucoupes Volantes,* Paris, 1977.
12. S.G. Soal and F. Bateman, *Modern Experiments in Telepathy,* Faber and Faber Limited, 1968.
13. Gary L. Blackwood and Daniel Cohen, *Paranormal Powers—Secrets of the Unexplained.*
14. P. J. Barber and D. Legge, *Perception and Information,* Methuen & Co., Ltd., London, 1976.
15. *Psychic Discoveries Behind the Iron Curtain,* Ostrander and Schroeder, 1970.
16. *Revista Saber Eletrônica,* many editions from 1976 to 1999, *Brazilian Electronics Magazine* (in Portuguese).
17. Newton C. Braga, *Som, Amplificadores e CIA* (in Portuguese), Editora Saber, 1995.
18. S. David Kahn, "Scoring Device," *Proc. Amer. Soc. for Psych. Res. Vol. XXV,* October, 1952.
19. John Fuller, *The Ghost of 29 Mega-cycles.*
20. Kroppner and Rubin, *The Kirlian Aura,* 1974.
21. W. Whately Carington, *The Measurement of Emotion,* New York, Harcourt, Brace and Co., 1922.
22. D. J. Ellis, *The Mediumship of the Tape Recorder.*
23. *The Secret Life of Plants,* Peter Tompkins and Christopher Bird, 1976.
24. Ken Webster, *The Vertical Plane.*
25. Jule M. D. Heisenbud, *The World of Ted Serios.*
26. Sonia Rinaldi, Transcomunicação Instrumental, *FE Editora Jornalistica,* 1997, São Paulo (in Portuguese).
27. Karl W. Goldstein, Transcomunicação Instrumental, *Editora Jornalistica FE,* São Paulo, 1992 (in Portuguese).
28. I. Chklovsky, *Univers-Vie-Raison,* ed., de La Paix, Moscow (in French).
29. Jacques Bergier, *Visa Por Une Autre Terre,* ed., Albin Michel, 1974 (in French).
30. Peter Bander, *Voice from the Tapes,* 1972.